Live Fish Transportation

NEW INDIA PUBLISHING AGENCY
New Delhi-110 034

About the Author

Dr. Dina Nath Pandit is presently working as an Associate Professor in the Department of Zoology, Veer Kunwar Singh University, Arrah – 802 301. He did his Masters in Zoology in First Class from TM Bhagalpur University, Bhagalpur – 812 007, India and Ph. D. from the same University in 2002 on a problem related to "Live Fish Transportation." He was the beneficiary of the ICAR and the CSIR fellowship during his research tenure. He won the Gold Medal from TM Bhagalpur University, Bhagalpur for securing First Class First in Master of Education in the year 1997. He taught undergraduate and postgraduate students for eighteen years besides guiding many researchers. He possesses experience in the teaching Animal Taxonomy, Fish and Fisheries, Biostatistics, Bio-techniques, Cell and Molecular Biology, Genomics, Biochemistry, Ethology and Environmental Biology. He has published around fifty-five research papers in the Journals of National and International repute. He has published two popular articles in daily news papers. He has presented paper in the Science Communicators' Meet during 103rd Indian Science Congress from January 3rd to 7th, 2016 at Mysore University, Mysuru. Earlier, he authored a textbook entitled "Animal Taxonomy: Principles and Practices."

He is a life member of the Society of Fisheries Technologists (India), Cochin and Indian Fisheries Association, Versova, Mumbai. Recently, He is elected as a member of the Royal Society of Biology's Council, London and it signify the he is a well qualified professional biologist subject to a rigorous code of conduct.

Live Fish Transportation

Dina Nath Pandit M Sc, Ph D, MRSB
Associate Professor
Department of Zoology
Veer Kunwar Singh University
Arrah – 802 301, Bihar, India

NEW INDIA PUBLISHING AGENCY
New Delhi-110 034

NEW INDIA PUBLISHING AGENCY
101, Vikas Surya Plaza, CU Block, LSC Market
Pitam Pura, New Delhi – 110 034, India
Email: info@nipabooks.com
Web: www.nipabooks.com

For customer assistance, please contact
Phone: + 91-11-27 34 17 17 Fax: + 91-11- 27 34 16 16
E-Mail: feedbacks@nipabooks.com

ISBN: 978-93-90591-69-5

Composed and Designed by NIPA

ICAR-Central Agroforestry Research Institute
Jhansi 284003, Uttar Pradesh, India

Foreword

The twilight of the 21st century saw the era of understanding of living organisms, preservation of nature and technical application of biological processes for the welfare of human beings. In this context, Fish and Fisheries encompass the knowledge of past, present and future trends, and their sustainability where the interaction of nature and human society is prevalent. The palatability and entertainment value of fish was recognized by the Roman Scholar Marcus Terentius Varro (116-127 BC). Since the dawn of civilization, it has been fascinating to accumulate literature on Fish, and several textbooks/field and laboratory manuals have been authored for the benefit of animal science in general and fish in particular. This important component of fish culture has been part of our age-old farming systems in the home yard leaving imprints of traditional agroforestry that provides integration of various enterprises. Of late, the socio-economic aspects of fish and fisheries have become so alluring that the UGC has introduced vocational courses like 'Industrial Fisheries' at the UG level and several Universities in India have come up with that.

An effort to write down a book on the domain of Fish and Fisheries for the students of M.Sc. in Zoology, B.F.Sc., M.F.Sc., Masters in Aquaculture and Fisheries, for fresh aspirants preparing for UGC-NET, CSIR-NET and a handful of other Examinations, as well as Researchers working in this field, is a matter of challenging journey as results of Fish and Fisheries has undergone radical modifications both in contents and implications. This textbook attempt to contain all the information including the latest one accumulated throughout the last two decades including the COVID-19 period related to Live Fish Transportation.

The book reflects that the author has had the option to collect entirely unexpected elements of Live Fish Transportation in lucid methods and organized their materials into a guideline back home-talk. Our syllabi, in a large portion of the Universities, even these days experience the ill effects of some of the genuine restrictions. The contemporary standpoint of Fish and Fisheries is not reflected in our educational programmes. In any case, inside the accessible system of the present–day educational plans, the author, Dr.

Dina Nath Pandit, Associate Professor, Department of Zoology, Veer Kunwar Singh University, Arrah – 802 301 has taken incredible endeavour to pass on the message of the Live Fish Transportation. It will certainly be beneficial to Fish traders, Fish raisers, Fish farmers as well as Aqua-culturists by this latest technology of live fish transportation. Eventually, it is expected that the technology will help to raise the GDP of the country as a whole. I strongly believe that this textbook will surely produce an effect on the modernization of the subject per se.

(A. ARUNACHALAM)

Director ICAR-Central Agroforestry Research Institute
Jhansi 284003, Uttar Pradesh, India

Acknowledgements

My first and foremost sincere acknowledgment goes to Dr. A Arunachalam, Director, ICAR-Central Agroforestry Research Institute, Jhansi, Uttar Pradesh – 284 003, India, for writing Foreword and giving brilliant thoughts as well as real insight to the book.

I extend my thanks to Prof. T.K. Ghosh, Department of Zoology, TM Bhagalpur University, Bhagalpur for his constant guidance, constructive criticism.

I acknowledge the regular help extended by Prof. J.S. Datta Munshi, Prof. S.P. Sinha, Prof. J. Ojha, Prof. J. N. Singh, Prof. S. P. Roy, Prof. Amita Moitra, Prof. M.C. Verma, Prof. P.K. Roy and Dr. D.N. Chowdhary 'Gopal' of Department of Zoology, TM Bhagalpur University, Bhagalpur as well. Whenever required, they always spared time from their busy schedule and giving me to bring this wonderful workout.

It's my proud privilege to pay my gratitude and regard to Dr. Anil Kumar Sinha, Prof and Head, Department of Zoology and Dean, Faculty of Science, Veer Kunwar Singh University, Arrah for his blessings and good wishes and also for providing me all the bits of help as well as valuable suggestions for this work.

The author extends special thanks to his colleagues Dr. Kizar Ahmed Sumon, Associate Professor, Department of Fisheries Management, Bangladesh Agricultural University, Mymensingh-2202, Bangladesh, Ilham Zulfahmi of Department of Biology, Faculty of Science and Technology, Ar-Raniry State Islamic University, Koperma Darussalam, Banda Aceh City-23111, Indonesia and Dr. Muhammad Akram (Associate Professor) Chairperson, Department of Eastern Medicine Government College University, Faisalabad, Pakistan for their valuable suggestion which has improved the quality of the book.

The kind support of Dr. Phillip Robinson, Associate Professor and Head (Coordinator, DBT Star Scheme and DBT PG Programme sponsored by DST and DBT, New Delhi), Department of Biotechnology, K.S. Rangasamy College of Technology, Tiruchengode, Tamil Nadu is sincerely acknowledged. His constant supervision and availability all the time are phenomenal.

I shall fail in my duty if I do not acknowledge the endeavours of Dr. Mohita Sardana and Dr. Amit Priyadarshi of Department of Zoology, Veer Kunwar Singh University, Arrah. Their efforts towards materializing this book shall always remain in my heart. I shall always fondly cherish those unfailing commitments toward their vision of the academic affluence among the teaching-learning fraternity.

The efforts made by Dr. G. R. Dutta and Dr. Dilip Thakur, Department of Zoology, TM Bhagalpur University, Bhagalpur to provide and revise the relevant content have to be acknowledged as well. His readiness to take this responsibility as a challenge is indeed worth acknowledging.

I am highly thankful to my all research colleagues. Without the help of Dr. K.K. Jha, Dr. S. Adhikari, Dr. A, Sengupta and Dr. R. Kumari the work would not be perhaps taken its present shape.

In this series of thanksgiving, I tender my sincerest thanks to all my family members and all my dears too. Their support, warmth and affection even beyond the day hours cannot go unacknowledged.

I extend my thanks to the authorities of Regional Sophisticated Instrumentation Centre (RSIC), Bose Institute, Kolkata and University Scientific and Instrumentation Centre (USIC), Jadavpur University, Jadavpur for providing Scanning Electron Microscope facilities.

I gratefully acknowledge the financial support provided by the Indian Council of Agricultural Research (ICAR) (No. 4(5)/89-ASR-I), New Delhi.

I tender my supreme thanks to our publisher, New India Publishing Agency, Pitam Pura, New Delhi – 110 034. Without their support bringing out this book would not have been possible. Their team was always at my disposal, whenever required. I convey my sincere thanks to them too.

Last but not the least, I am highly thankful to my kid, Ms. Deepa Sonal, Assistant Professor, Department of Computer Science, Patna Women's College, Patna for sparing time in this dedicated work.

(Dina Nath Pandit)

Preface

Fish is promoted as the most advantageous of creature proteins for people. Eating fish in any event is suggested by all driving wellbeing bodies for insurance against heart sicknesses, stroke, diabetes, pulse etc. Fish has been a part of conventional Indian weight control plans of the beach front locales for quite a long time. It is anything but difficult to source and cook and can be singed, bubbled, heated or broiled. It packs incredible flavor alongside medical advantages. Indian eating regimens are especially inadequate in proteins besides in zones where fish is effectively accessible.

Pisciculture has received increased attention in various parts of the globe to increase food resources. The issue of transport of live fish or their seeds is closely linked to a successful pisciculture programme. Modern methods of transporting live fishes are practically difficult in India, where communication is insufficient and distance prohibitive. Air-breathing fish culture attracted the attention of aqua-culturists during late 20th century, and many varieties of air-breathing fishes were successfully bred using heteroplastic pituitary extract (Ramaswamy and Sundararaj, 1957; Moitra et al, 1979). However, because of their heavy mortality, traditional transport of spawn, fry and fingerlings in different open and closed containers poses major problems. Different factors such as oxygen depletion and increase in carbon dioxide tension, ammonia accumulation, pH reduction, overcrowding injury and hyperactivity (Jhingran, 1991, Pandit and Ghosh, 2012) are responsible for the death of different fish stages. Mortality during the transport of live fish is high, mainly due to stress. Stress tolerance levels of some commonly cultured fish are given below:

1. High tolerance levels are shown by black carp, catfish, common carp, mud carp, gourami and *Oreochromis.*
2. Medium tolerance levels are shown by bighead carp, Indian carps and grass carp.
3. Low tolerance level is shown by silver carp.

India is home to in excess of 10% of the worldwide fish variety. As of now, the nation positions second on the planet in complete fish creation with a yearly fish creation of about 9.06 million metric tons. Freshwater fishery adds to more than 95 percent of the total aquaculture production.

Very recently, COVID-19 Coronavirus and the ensuing lockdown have adversely affected Indian fisheries. In any case, India isn't the only one. Extreme monetary misfortunes to fisheries have been accounted for across the world, and in all cases, the harm has been most noteworthy for send out driven fisheries with complex stock chains (Stokesa et al, 2020).

Resumed fisheries confronted disturbed ware chains, a huge absence of customer interest, and a rising apprehension of being tainted by COVID-19. Transient fishers—who make up an enormous piece of India's automated fishing labour force—were in no condition to go fishing. They had been abandoned for quite a long time in unfortunate conditions, on boats or at swarmed fishing harbors and they just wished to return home. Every one of these circumstances is showing the significance of a superior transportation strategy to vitalize fishery area.

The oxygen requirement of a fish depends upon a variety of extrinsic and intrinsic factors that play an important role in body size. It is essential to assess the oxygen requirement of these fishes in order to overcome the problem of high mortality of fishes during transportation at their early stages of development.

Fish are generally transported in containers such as cans of different sizes, pots of ceramic or metal, wooden or metal buckets, vats, barrels, plastic bags, Styrofoam boxes, bottles, jugs, animal skins and bamboo sections. Almost any clean, waterproof container may be used. Certain containers provide good insulation from heat, for example, wood or Styrofoam. Containers like metal or plastic are poor insulators and may have to be wrapped with wet towels or packed with ice to keep temperatures down.

In India, various species of *Anabas, Channa* (=*Ophiocephalus*), *Clarias, Heteropneustes* (=*Saccobranchus*) and *Osphronemus* are considered to be economically important air-breathing fishes in Bihar, Bengal, Uttar Pradesh, Assam and South India's swamps and weedy water. They are known for their higher nutritional, recuperative and medicinal characteristics, which sometimes preferred over major carps and recommended during convalescence as a diet.

The interest for live fish is high in India, and its cost surpasses from the frosted dead fish by INR 100–250/kg. The Indian fish market was worth INR 1,110 Billion during 2018. The market is projected to arrive at INR 1,998 Billion by 2024.

Anabas, commonly known as the climbing perch, koi or kawai belongs to the order Perciforms. It is widely spread across countries in South East Asia including India. *Anabas* has narrow trunked species as *Anabas testudineus*

(Bloch). It is cultivated with *Heteropneustes* (=*Saccobranchus*) *fossilis* and *Clarias batrachus* and grows to about 15 to 23 cm. Fry and larval stages of *Anabas testudineus* are zooplankton feeders, but they become carnivorous later.

Four species of *Channa* (Order: Ophiocephaliformis) are common in India, namely, *Channa punctatus* (Bloch, 1793) or *Channa punctatus* (Menon, 1974), *Channa marulius* (Hamilton, 1822), *Channa gachua* (Hamilton, 1822) and *Channa striata* (Bloch, 1793). Two new species of *Channa* namely *Channa aurantipectoralis* and *Channa pardalis* have been reported recently (http://www.catalogueoflife.org/ Accessed 2017). They are also known as snake headed murrel or garai and regarded as a good table fish in several parts of India. *Channa punctatus* matures when they are about a year of age (body weight=8g and total length=about 8.5cm). Progress in murrel culture depends upon the success in raising the fry and fingerlings.

The developmental stages of Indian air breathing fishes can be shipped without oxygen pressing. Polythene drums or iron drums are utilized for transport of fry or fingerlings. The transporter should have enough space for their routine surfacing to breathe climatic air.

The essential standards of fish transport and the principle factors influencing it (fish species, developmental stages of fish and quality, length of transportation, temperature, oxygen content, group size, behavior of fish, fish impact of anaesthetics, digestion items, and so on) will assessed based on an examination of the appropriate writing. The shut vehicle units will be portrayed and the densities of moved fish per unit volume under genuine conditions will be organized for direction.

It was observed that many technological and conditional changes, like pandemic due to COVID-19, have been reported at both the local and global level concerning live fish transportation. In this context, the book is written with a hope that it will invigorate the interest among the students as well as the teachers. It covers all the important aspects of live fish transportation along with the latest related developments and information. Proper sanitization of all the equipments and accessories is one of the essential steps for fish transportation in this pandemic.

The book has been compiled in ten chapters. Chapter-III to Chapter-VII is sub headed into Introduction, Materials and Methods, Observations and Results and Discussion. Subject index is designed to locate the important terms and name of fish. A separate author index has also been added. The cited references are enlisted under bibliography. The summary of the contents of the book may be pointed as:

Chapter-I is planned to discuss introductory aspects of fish transportation.

Chapter-II explains brief historical review of the process of fish transportation.

Chapter-III begins with the process of determination of aquatic and bimodal oxygen uptake of fishes. The detailed account of oxygen uptake seems to be plinth for fish transportation.

Chapter-IV gives an account of oxygen uptake of fishes under association in still water and continuously flowing water conditions. The information is necessary for the calculation of either increase or decrease of oxygen requirement during packing of fishes in a grouped condition. The study helps in determination of optimum number of fishes considered for packing.

Chapter-V explains determination of optimum dose of anaesthetics namely Benzocaine and MS-222 of fishes and its effect on the oxygen uptake. The information is helpful for the calculation of decrease of oxygen requirement during packing of fishes.

Chapter-VI deals with effect of ammonia on ultra structure of gill under scanning electron microscope and oxygen uptake of fishes. The study is helpful in deciding calculation of complete requirement of oxygen for packing.

Chapter-VII critically explains evaluation of suitable packing conditions based on the information from Chapters III to VI under normal situation and present situation of COVID-19.

Chapter-VIII reviews a general discussion of entire account of various processes of fish transportation.

Chapter-IX describes summary and conclusions observed from the work of fish transportation.

Chapter-X deals with a flow sheet, a proposal and a future scope of fish transportation.

I believe that the present book will be useful for persons and organizations pursuing studies concerning fish transportation of either carps, air-breathing and other fish. It will be also helpful for the aqua-culturists taking up culture and research of fishes in different parts of India and abroad. Although, care has been taken during the preparation of the book but the shortcomings are inevitable. I, therefore welcome suggestions, valuable comments and constructive criticism.

Author

Contents

1

General Introduction

Inland fisheries are important global contributors to food security, livelihoods, and ecosystem services. A report on the global snapshot of expert-perceived impacts of COVID-19 on inland fisheries has been submitted by Stokesa et al, (2020). They studied an online survey to determine perceptions of Inland fisheries pressures between June and July 2020. They also analysed pandemic impacts on inland fisheries, based on the responses analyzed against fish capture and human development index data.

(1) Thus promoting a sustainable alternative livelihood and social buffer with low barriers to entry.

(2) In rapid crises.

(3) Tentative evidence of anthropause.

(4) Some positive ecological reactions and better functioning of an ecosystem has been suggested as a consequence of the COVID-19 pandemic and

(5) The ecological responses of some resources may differ based on how the resources are used (e.g., fish for food or leisure).

Inland fisheries may deliver short-term, mid-pandemic jobs (e.g., fishing and processing), food supplies and socially distanced leisure, but the sustainable yields are also threatened by changing harvest constraints and market demands aggravated or amplified by preexisting pressures (e.g., dams and pollution). The study assessed expert-perceived pressures of COVID-19 on global inland fisheries and relates these data to three usage indicators:

1) Reported inland fish catch,
2) Human development index [HDI; composite metric: life expectancy, education, and living standard (5)] and
3) Fishery provisioning value.

Promoting live fish is viewed as a worth adding methodology because live fish get significantly greater costs contrasted and new chilled or frozen items. The interest for live fish is high in India, and in light of this, the cost surpasses that for frosted dead fish by INR 100–250/kg. The Indian fish market was worth INR 1,110 Billion during 2018. The markct is projected to arrive at INR 1,998 Billion by 2024.

Transportation of attractive and marketable size and brood fishes faces various issues. Of late, the expanding prominence of aquaculture and consumer inclination for live fishes has pulled in the consideration of aqua-producers and fish brokers. Consequently, transporting fishes in trucks mounted with compartments, aluminum hundies, cylindrical and hollow high-thickness polyethylene tanks and in some cases canvas rolled containers are being utilized for this reason. Notwithstanding, these transportation strategies for large size fishes had certain constraints and the absence of appropriate logical reinforcement concerning oxygen utilization contemplates.

Fish health is influenced by changes in water quality parameters while in the plastic bags during transportation. The parameters to be considered are temperature, dissolved oxygen, pH, carbon dioxide, ammonia, anaesthetics and the salt equilibrium of the fish's blood. The pace of progress of every parameter is influenced by the weight and size of fish to be transported and the duration of transport.

Air-breathing fish are often characterized by a significant reduction in gill surface area to avoid loss of oxygen taken up in the air-breathing organ (ABO) at the site of the gills in hypoxic water (Pandit and Ghosh, 2005; 2007; 2012; Scott et al, 2017; Pelster et al, 2020).

Respiration of air-breathing fishes has interested researchers for at least a century; the earliest attempts to measure oxygen uptake ($\dot{V}O_2$) and respiratory partitioning date to the 19th century (Baumert, 1853). The available general respirometry methods have been described and reviewed by several authors (Cech, 1990; Lighton, 2008; Lefevre et al, 2016; Zhang et al, 2019) and studies of the respiration of air-breathing fishes were reviewed in detail in the iconic book by Graham (1997).

Oxygen consumption of fish reflects its basal metabolic status and is one of the indicators of the general health or/and well being of the fish. It may also be useful to assess the physiological state of an organism, helps in evaluating the susceptibility or resistance potentiality and also useful to correlate the behaviour of the animal, which ultimately serve as predictors of functional disruptions of population. Analysis of oxygen consumption can be used as a bio detector system to evaluate the basic damage inflicted on the animal which could either increase or decrease the oxygen uptake (Neelima et al, 2016).

Oxygen is the prime requirement of all living beings and fish is no exception. Due to the lack of adequate oxygen present in the water, high mortality is noticed during the transportation of live fishes (Jhingaran, 1985). Knowledge of the oxygen consumption rate is a prerequisite for the determination of the species–specific density of fish size, duration of transport and water requirements during live transportation. Transportation of live fishes in large quantities is one of the major challenges for aquaculture and even for export due to the lack of adequate transportation equipment. The use of oxygen-filled polyethylene bag is in vogue for the transportation of fish seed.

Oxygen utilization paces of spawn, fry and fingerlings of carp have been concentrated by numerous specialists (Basu, 1951; Hora and Pillay, 1962; Ramchandran, 1969; Singh, 1977; Selvaraj et al, 1981; Bhaumik et al, 1986). Fry, fingerlings, adolescents and table fishes are shipped live in different sorts of conventional containers by sprinkling and circulating water inside the compartment (Jhingaran, 1975; Jhingaran and Pullin, 1985; Ou and Li, 1996). There have been restricted endeavours for the transport of live fishes by different strategies, especially brood fishes with a certain level of achievement (Mammen, 1962; Patro, 1968; Ramchandran, 1972; Das and Bhaumik, 1979).

The fish to be transported should be healthy and in fitness. Weakened specimens ought to be eliminated from the consignment, significantly once the temperature throughout the cargo is high. Once the fish are of poor quality, even a good reduction of fish density within the transport instrumentality fails to forestall fish losses. Weakfish are killed at a way higher rate than fish in fitness once the transport time is longer.

The greatest load of fish that can be safely transported inside a given timeframe is the carrying capacity. The carrying capacity relies upon the length of haul, water temperature, fish size and fish species. If water quality conditions, for example, temperature, oxygen, carbon dioxide, alkalinity and ammonia are consistent, at that point carrying capacity will rely upon the fish species. All in all, fewer pounds of more modest fish can be moved per gallon of water than bigger fish. It is significant that first-time or experienced transporters dealing with another species trial a bunch before undertaking a huge shipment.

Days before the fish are loaded and transported; the transporter ought to decide the transporter to be utilized, the season of takeoff, the season of appearance and transportation costs. This data should be conveyed to the beneficiary well before the transportation date. All stacking should be intended to permit boxes to be delivered not long after stacking as could reasonably be expected. With appropriate arranging, superfluous postponements in conveyance and pickup can be evaded. The recipient is liable for reaching the transporter if any deaths happen.

Among freshwater edible fishes, murrels *Channa* species, perch *Anabas testudineus* fetch high market demand followed by catfishes by consumers of South and South-east Asia but their culture systems are yet to be established in many Asian countries. As far as freshwater air-breathing catfish are concerned, the stinging catfish *Hetropneustes fossilis* is very common and high priced due to tender flesh, delicious taste, less fat, high digestibility and medicinal value and survival in oxygen-depleted water, resistance to overcrowded stress and acceptance of pellets.

There are two/three established transport systems for live fish. The choice of this transportation system depends on the quality and quantity of the transported species, transportation facility available, transportation duration, size of market etc.

(a) The closed system: The closed system could be a sealed container during which all the requirements for survival are self-contained. The best of those could be a sealed bag partially full of water and oxygen.

Bag shipment might be the most ideal decision for the transporter for a few reasons. To start with, little fish and fry could be harmed by being dispatched in huge tanks. Second, because of the extraordinary distances included, bag shipment may offer financial preferences over standard tank transportation. This reality sheet will zero in on the vehicle of fish. With minor alterations, the procedures and chiefs talked about additionally apply to fish.

The regular shape of bags sometimes has dimensions of 0.8–1.1×0.35–0.45m. Usually, the upper end is typically fully open. The bottom either contains a seam within the middle or consists of an oblong piece of foil; the latter variant is better as a result of which helps to prevent loss of the fish caught within the corners.

During transport, fries are used in outer cases to protect the bags from mechanical damage, in particular by punching or tearing in contact with the ground. The case holds the bags in the desired location, enables storage and/or provides thermal insulation for the bags.

These cases may include carton boxes, appropriate plastic containers, big polyethylene cans, large polystyrene boxes. The type of outer housing depends on the number of bags transported, duration and methodology of transport, specifications for further handling and on the differences between ambient temperature and also the temperature of the water within the container.

When large fish species are transported for introduction or acclimatization, it is not recommended to use the theoretical vital possibilities. The main requirement is to stay the fish healthy and physiologically intact throughout the transport as a result of brood fish are of a high potential value and their transport is to pay off within the future. For this, the safe transport densities of these fish, as recommended by Orlov et al, (1974) are 5 to 10 times lower than those of the fish of a similar size transported to market.

(b) The open system: The open system consists of water-filled containers during which the requirements for survival are supplied continuously from outside sources. The simplest of these could be a small tank with an aerator stone.

In all cases of fish transport in open systems, it ought to be borne in mind that even a short-time transport of 10–30 m in open plastic or metal tanks ought to be done underneath the conditions of constant air or oxygen supply. This may be very important to the welfare of fish even though the dissolved oxygen content of water appears to be satisfactorily high within the tank. Transport longer than half an hour ought to be incompletely filled and closed tanks to prevent splashing and injuries to young fish bumping into each other in the well of the tank.

In some countries, the cargo bed of a 10-ton truck is covered with canvas so it can hold water. The water volume depends upon the area and profundity of the bed and the conveying limit of the truck in the "open truck system". On landing in business sectors, fish overseers enter the water in the truck to get the fish and move them to the wholesalers. This can cause COVID-19 transmission when fish are given over between laborers.

The COVID-19 is spread by droplets from coughing, sniffling or talking. The droplets are either breathed in or advance into the body through contact with the eyes, mouth or nose. What's more, the settled droplets can stay dynamic on surfaces for quite a while, which could debase others. That is the reason there is a dire need to change the conventional working practices in the live fish transportation area. There are numerous elements to mull over for effectively shipping live fish. Improving these could bring about the lower transmission of COVID-19 among laborers and less pressure and mortality for fish.

(c) Altered waterless framework with no water, aside from being kept damped by utilizing pre-chilled sawdust or wood shavings.

Data on the potential densities of fish transported in open systems are scarce in the literature. Although Leis (1978) delineated a calculation methodology for the oxygen requirement in fish transport, his methodology cannot be widely used in practice because the calculation formulae are very sophisticated. On the other hand, many authors stated that fish consume only a small portion of the oxygen offered (10 percent at the maximum, as asserted by Proske, 1982), so that the calculations of oxygen requirement, based on the amount consumed by the fish, are not reliable. Like in the closed systems, it is conjointly true that the optimum fish densities in transport containers are influenced by many factors that ought to be learned and evaluated by practical experience rather than by theoretical calculation. Most of the authors also recommend taking into consideration the possibility of delay throughout the journey which may be up to 24 hours in longer transports.

Some farmers and traders use inappropriate transportation practices, which have led to health hazards during the COVID-19 pandemic. Besides, poor transportation techniques have resulted in high mortalities during shipping. Among those that do survive and make it to the market, many arrive in poor quality because of injuries sustained on the trip. On top of all this, these hazardous techniques also have dangerous impacts on road safety.

These systems are reviewed and delineated in this book in conjunction with some new information and techniques for the issues of fish preparation for transport, types of vehicles and equipment, issues of water quality and its changes throughout transport and chemical aids used during fish transport.

This book gives details of different essential aspects of live fish transport in these three principle regions:

1. Capture and pre-transport maintenance: The capture and pre-transport maintenance of live fish involves some modification of techniques commonly used by commercial fishers and processors. Detailed information is given on the capture and handling techniques, the construction and maintenance of live fishing facilities, and design and operation of recirculating filtration systems.

2. Packaging and live transport: There are fundamentally 3 strategies that are commonly used for transporting live finfish by air:

(a) The polystyrene seafood box;

(b) The pickle barrel system and

(c) The big box system.

The auxiliary goals of this book are to improve human wellbeing during the COVID-19 pandemic comparable to reaping and moving live fish. Different targets incorporate

(i) Providing more secure techniques for dealing with fish transportation,

(ii) Improving creature government assistance,

(iii) Reducing street security risks and

(iv) Detailing techniques to lessen the expense of live fish transportation.

3. Post-transport upkeep: The wellbeing of shipped fish was assessed regarding both fish wellbeing (bacterial and ectoparasite levels) and levels of potential human wellbeing microorganisms (*Salmonella almonellae* and *Vibrio parahaemolyticus*).

The pressing cycle may require a few hours, contingent upon which bundling framework is utilized. Most aircraft require airship cargo to be conveyed 2-3 hours before the planned takeoff time. At the objective, the cargo should be eliminated from the airplane, and on account of global ports, be cleared by Customs. The compartments should then be conveyed to the holding office before they can be unloaded, a cycle that may require a few hours.

Live Fish Market Status

Globally live fish trade is well known, especially in most of the South-East Asian and the Southern Pacific regions. In India, initially live fish transportation was limited to the North-eastern states, however given the increased demand for live seafood commodity, currently, there is a highly lucrative market for live fish. However, in the Indian scenario, live trade is mainly limited to carps as well as air-breathing hardy fishes such as catfishes. Live fish shipping methods include the use of primitive techniques such as transportation in conventional baskets made of bamboo or coconut coir lined with leaves to large plastic containers or aluminum vessels for short-distance transportation. Insulated or non-insulated containers with ice or cooling gels to large trucks are often used for local transportation. Hauling trucks with more sophisticated equipment such as assembly for filtration, oxygen generation, refrigeration and reticulation of water are used for large scale operations involving long-distance transport of high-value seafood but are very costly.

All these are open transportation systems wherein facilities for aeration, recirculation etc. are supplied from external sources.

2

Historical Resume

India is facing with the issue of food production, especially the production of protein resources. The food production rate in India is insufficient to cope up with an alarmingly growing population. Protein deficiency is currently the most serious human malnutritional problem, with protein deficiency affecting about 30% of the world's population. To cope up with the current situation, there is therefore a need for an all-out effort to increase food production especially the production of animal protein. Pisciculture has acquired primary importance in this context, and the culture of Indian major carps and their allies is now a common practice. For a long time, the culture of air-breathing fishes in the derelict water has remained neglected. Many such fishes, however, have become popular in the recent past, being rich in their food value.

The demand for fish as nutritious food is increasing day by day, as it is a good source of high quality-protein and lipid and is more palatable than vegetables. It is relished for its characteristic flavour due to high myofibrillar and low stroma protein, rich in essential amino acids particularly lysine and arginine (Kinsella, 1986). Fish proteins play a beneficial role in lipid and cholesterol metabolism by reducing triacylglycerol (=triglyceride) and phospholipid levels in plasma (Tsukuda and Arai, 1986). These proteins occupy an important place in human nutrition due to their high digestibility and growth-promoting value. Fish lipid has a high amount of polyunsaturated fatty acids (PUSFAS), which reduce plasma cholesterol, and this helps in ischemic heart disease and arteriosclerosis (Kinsella, 1986).

Indication of the use of fish as food has been obtained from the excavation of Mohenjodero and Harappa of the Indus Valley Civilization some 2500-1500 BC back (Nath, 1966). Aristotle (384-327 BC) is regarded as the founder of Ichthyology but Somesvana first recorded the common sport fish of India (Hora, 1951). The writings of Kautilya in Arthashastra indicate that pisciculture probably flourished in reservoirs (Hora and Pillay, 1962). From ancient times, man has utilized ponds and kept fishes for their interest. The knowledge of the occurrence of fish in India is some three millenniums BC old (Hora, 1956). In this country, pisciculture was first started in West Bengal about 1500 years ago and gradually spread to the other parts of the country.

The Government is now taking vigorous steps for the conservation and development of their vast fisheries resources. Research conducted by Central Inland Fisheries Research Institute (CIFRI) has revolutionized fish culture in India. India's bounty of natural aquatic resources inhabits about 1479 species of fish (Das, 1990), of which about 296 species are economically important. In this concern, the cultivation of many economically useful fish, artificial fertilization, induced breeding, etc. has opened a new era in the pisciculture programme. Thus, the fishery is useful in the production of food, contribution to national income, employment opportunity, recreation and export etc. In other words, fisheries occupy a major position in the national economy and their role is increasing day by day. Considerable progress has been made to increase per hectare yield by controlled breeding, transportation and rearing of young ones. Attention has also been given to improve transportation techniques because the success of pisciculture largely depends on the safe transport of the fishes or their seed. Central Inland Fisheries Research Institute has contributed significantly to the development of fishery in India but the progress has mainly been achieved in the culture of carps and their allies.

The harvesting team uses drying as a means of disinfecting the nets, but even then only if time and weather conditions permit. These practices run the risk of transmitting fish diseases, particularly on rainy days. Since the COVID-19 virus may survive in water, there is possibility of human infection from freshly used wet nets (Anon, 2020).

In India, various types of air-breathing fishes are also found. Of these, *Anabas testitudineus, Channa punctatus, Clarias batrachus and Heteropneustes (= Saccobranchus) fossilis* etc. are some of the economically important air-breathing fishes. They usually show fast growth during their early life as compared to herbivorous species (Jhingran, 1991). Accordingly, the culture of air-breathing fishes either by monoculture or polyculture has been started by the All India Coordinated Project on Air-Breathing Fishes under the Central Inland Fisheries Research Institute at Bhadra Centre (Karnataka) and Darbhanga Centre (Bihar) to serve as a source of protein especially for the rural people. Some air-breathing fishes are treated as excellent table fish in many states of India, and their transportation to distant places in live conditions poses a problem. Although, different aspects of respiratory physiology of these fishes have been studied but studies concerning their physiological demand during transportation etc. have not been properly attended to.

Oxygen Uptake Concerning Body Weight

A fish normally utilises oxygen from the water it lives in and a decrease in oxygen level leads to unfavourable conditions for the fish and may result in their mortality. The oxygen uptake in fishes is taken as a measure of metabolism and is directly related to their body weight. The relationship between body weight and oxygen uptake is an essential and useful prerequisite for packing, transport stocking and rearing etc.

Rübner (1883) was probably the first person to study the relationship between oxygen uptake and body weight in dogs and concluded that metabolism is proportional to the 2/3rd power of body weight. His findings were later corroborated by Morgulius (1914). A slope of 0.9 was calculated by Smith (1935 a and b) to show the relationship between body weight and metabolism of fishes. A lower slope of more or less 0.5 was suggested by Wells (1935), Barlow (1961), Ruhland (1965) and Morris (1967) in their works. Zeuthen (1947) reported a slope of 0.78 for different species of fishes. Scholander et al, (1953); Job (1955); Winberg (1956); Hemingsen (1960); Prigogine and Wiame (1964); Brett (1965); Paloheimo and Dickie (1965) and Prama de Croux (1997) reported slopes between 0.75 and 0.85. Saunders (1963) suggested a slope from 0.74 to 0.84 in Atlantic cod to show the relationship between body weight and oxygen uptake. Kamler (1972) reported a slope of 0.78 and 0.80 for smaller and larger fishes respectively to show intraspecific variation due to growth. Khamleva (1973) on *Idotea balticabastin* estimated a slope of about 0.70 to express the relationship between body weight and oxygen uptake. A similar slope value was also obtained by Ojha and Munshi (1975) while working with *Macrognathus aculeatum*. Jager and Dekkers (1975) suggested a slope of 0.82 for the relationship between oxygen uptake and body weight. Wares II and Igram (1979) calculated a slope of 1.053 to establish the relationship between body weight and oxygen uptake in fathead minnow, *Pimephales promelas*. In the recent past, Prama de Croux (1997) recorded routine oxygen uptake of *Salminus maxillosus* and calculated a slope of 0.70.

Besides these, information is also available on different aspects of aquatic and bimodal oxygen uptake of air-breathing fishes. Aquatic oxygen uptake of Indian air-breathing fishes, using cylindrical glass respirometer with continuously flowing water with surfacing allowed (access to air) and surfacing prevented (without access to air) conditions is known from the experiments of Munshi and Dube (1973) on *Anabas testudineus*; Munshi et al, (1976) on *Clarias batrachus* or magur; Ojha et al, (1977) on *Colisa fasciatus*; Munshi et al, (1978) on *Heteropneustes fossilis* or singhi; Ojha et al, (1978) on *Channa gachua* or gurai; Ojha and Singh (1981) on *Mystus cavasius* or tengra; Hakim et al, (1983) on *Channa punctatus* or snake headed murrel; Singh and Munshi (1985) on *Glossogobius giuris* and Ojha and Haque (1994) on *Pseudapocryptes lanceolatus*, etc.

Oxygen uptake through bimodal gas exchange system has been studied in a few Indian air-breathing fishes such as *Anabas testudineus* or kawai (Hughes and Singh, 1970; Singh. 1976), *Heteropneustes fossilis* (Hughes and Singh, 1971), *Clarias batrachus* or magur (Singh and Hughes, 1971), *Amphipnous cuchia* (Lamholt and Johansen, 1976), *Channa marulius* or gurai (Ojha et al, 1979) and *Notopterus chitala* or chital (Ghosh et al, 1986) etc.

The relationship between bimodal oxygen uptake with body weight among Indian air-breathing fishes has been worked out by Munshi et al, (1976) on *Clarias batrachus*; Munshi et al, (1978) on *Heteropneustes fossilis*; Ojha et al, (1978) on *Channa gachua*; Choudhary (1979) on *Channa striata*; Biswas et al, (1979) on *Boleophthalmus boddaerti*; Ghosh et al, (1980) on *Monopterus cuchia*; and Ojha et al, (1988) on *Colisa* (=*Trichogaster*) *fasciatus* etc.

The bimodal oxygen uptake concerning body weight at different seasonal temperatures in various air-breathing fishes has been investigated viz., *Amphipnous cuchia* (Lamholt and Johansen; 1976), *Heteropneustes fossilis* (Munshi et al, 1982), *Clarias batrachus* (Patra et al, 1983), *Anabas testudineus* (Ghosh and Munshi; 1987) and *Channa punctatus* (Ghosh et al, 1990) etc.

The oxygen uptake in fishes is influenced by many abiotic and biotic factors, such as, pH (Wiebe et al, 1934; Packer and Dunson, 1970; Divelly et al, 1977; Bhasker and Govindappa, 1985; Munshi and Singh, 1992; etc), temperature (Schaeperclaus, 1933; Sumner and Lanham, 1942; Parvatheswararao, 1959; Prosser and Brown, 1961; Ojha and Munshi, 1975; Wares II and Igram, 1979; Mosser and Hettler, 1989; Barrionuevo and Fernandes; 1998 etc.), partial pressure of oxygen and carbon dioxide (Rahn, 1966; Johansen and Lenfant, 1967; Hughes and Singh, 1971; Wares II and Igram, 1979; Yamamoto, 1991 and 1992; Cruz-Neto and Steffensen, 1997; etc.); salinity (Farmer and Beamish, 1969; Mathew, 1976; Pandey, 1975, Mosser and Hettler, 1989; Jha and Ghosh, 1991; Pathak et al, 1992; Peterson-Curtis, 1997; etc.), pollutants (Pandey et al, 1976; Reddy et al, 1977; Singh and Singh, 1979; Roy and Munshi, 1988; Bhoopathy and Gunesegar, 1999; Thorat and Wagh, 2001; etc.), physiological condition or activity and exercise (Fry, 1947; Fry and Hart, 1948; Brett, 1965: Lipskaya, 1974; Crawford, 1978; Schurmann and Steffensen, 1997, etc.), hormones (Packard and Stevensen, 1974; Pandey and Munshi, 1976; Chanchal and Pandey, 1979; Jha, 1995 etc.), nutrition (Smith, 1935a and b; Beamish, 1964a and b; Edwards et al, 1971; Cech, 1990 etc.), rhythmic behaviour in oxygen uptake (Patra et al, 1978; Munshi et al, 1979; Ghosh and Biswas, 1980;

Ghosh et al, 1990 and Munshi and Ghosh, 1993; etc.), stages of life cycle (Lipskaya, 1974; Kamler, 1976; Munshi et al, 1978; Chanchal and Pandey, 1979, Jha, 1995 and Subba,1999 etc.) and inter or intraspecific variations (Barlow, 1961; Ojha et al, 1981; etc.).

Pandit et al, (2008) investigated oxygen uptake in *Channa punctatus* (Bloch) from pond water versus tap water. Pandit et al, (2009) also worked on comparison of oxygen uptake of *Channa punctatus* (Bloch) from tube-well water and tap water.

Effect of Group Size on Oxygen Uptake

Experiments on fish aggregation have shown that the oxygen uptake of a fish is affected by their number participating in respiration. The variation in oxygen uptake, when fishes of either similar or different genera, species or subspecies are kept together, is known as the group effect. The variation is probably related to the ethological aspect of fishes. Considerable work has been done by various workers on fish aggregation to establish the effect of group size on oxygen uptake.

Fishes show either soothing or calming behaviour, aggressive or excitatory behaviour, shoaling or schooling behaviour under association. Umezawa et al, (1983) suggested that the group effect is influenced due to the soothing or excitatory behaviour of the fish. A concept of shoaling behaviour in fishes was given by Kennedy and Pitcher (1975). Griogr'eva (1966) investigated the impact of schooling on the oxygen uptake of fishes. The excitatory or aggressive behaviour has been suggested to elevate the oxygen uptake (Umezawa et al, 1981).

Allee (1931) was probably the first person, who observed that oxygen uptake of a fish is reduced due to the group effect. While working with catfish (*Ameriurus melas*), Boven (1932) observed no significant difference in the oxygen uptake of isolated and grouped fishes, although the rate of oxygen uptake of isolated fish was higher. Schuett (1934) demonstrated that the locomotor activity of goldfish increases in isolation than when they are in a group of four. Similarly, Shlaifer (1938) found that an isolated goldfish (*Carassius auratus*) has more oxygen uptake and high locomotor activity than the fish in a group of two or four. According to Shlaifer (1939), the decrease in oxygen uptake due to aggregation in goldfish is lost under dark or in blinded conditions. Reduced oxygen uptake in a fish when it is in a group has also been reported by Geyer and Mann (1939), Job (1955), Delco (1963), Rhyzhkov (1970), Parker (1973) etc. The schooling effect to vision has been worked out by Malyukina et al., (1962) in minnow (*Phloxinus phloxinus*) and goldfish (*Carassius carasssius*).

Uematsu (1971) also found that although the rate of feeding was higher in the guppy, yet the rate of oxygen uptake and locomotor activity was lower under aggregation than in isolation. While working with fathead minnow, *Pimephales promelas*, Wares II and Igram (1979) explained that a reduction in oxygen uptake occurs in a group of two and three fishes only. They also suggested that aggregation probably diminishes motor activity. The effect of grouping on the oxygen uptake of medaka (*Oryzias latipes*) was investigated by Umezawa and Watanabe (1979) and they found that oxygen uptake of a fish under isolation was more than a fish in a group of ten, suggesting lowering of oxygen uptake in grouped condition. Similarly, a lowering or reduction in oxygen uptake during post-juvenile and larval stages of red sea bream (*Pagrus major*) has been observed due to group size by Imabayashi and Takahashi (1987).

Contrary to the above observations, Bowen (1932) found that catfish (*Ameriurs melas*) in a group of four have a greater rate of individual oxygen uptake than those in isolation. Brett and Sutherland (1965) investigated that the aggressive behaviour among pumpkinseed (*Lepomis gibbosus*) elevated oxygen uptake due to the excitatory effect. According to Wirtz and Davenport (1976) the sight of mirrored images in blennies (*Blennius pholi*) increases the oxygen uptake not only when the fish is placed in visual contact of conspecific fish but also with a heterospecific companion having similar habits. Umezawa and Kanikawa (1980) found that an isolated goby (*Tridentiger obscurus*) consumed less oxygen than a fish did under visual contact with other fishes living in nearly similar habitats

Umezawa and Watanabe (1979) measured oxygen uptake of medaka (*Oryzias latipes*) and found that oxygen uptake of isolated medaka is not modified much by visual contact with its mirror images or with its surrounding. Group effect on oxygen uptake and some observations on aggressive behaviour of goby (*Tridentiger obscurus*) were studied by Umezawa et al, (1981). Umezawa et al, (1983) investigated group effect on the oxygen uptake of aye (*Plecoglossus altivelis*) to explain variation in behaviour during different growth stages.

Information on oxygen uptake to group size on Indian fishes is limited (Singh and Munshi, 1985), although it's a utility in the transportation of fishes was considered by Motwani and Bose (1957). Jha (1995) investigated the effect of group size on the oxygen uptake of early developmental stages of two Indian major carps namely *Labeo rohita* (rohu) and *Catla catla* (Catla). Such information on the Indian air-breathing fishes is probably lacking excepting the work of Ghosh et al, (1994) on the air-breathing fish, *Heteropneustes fossilis*. A comparative study on oxygen uptake during still water aggregation of *Anabas testudineus* and *Channa punctatus* was conducted by Pandit and Ghosh (2006). Oxygen uptake to group size in the juveniles of, *Anabas testudineus* was investigated by Pandit and Ghosh (2007).

Effect of Anaesthetics on Oxygen Uptake

Handling of fishes may cause physical damage to the fish, loss of mucilaginous covering which makes the fish susceptible to pathogenic attack, stress and haematological changes etc. The use of anaesthesia reduces the chance of physical damage and provides a stress-free condition to the fish. Anaesthetics are soothing medicines, which sedate the fish and consequently decrease oxygen uptake, removal of carbon dioxide and ammonia etc. It also controls the excitability of the fish and thereby reduces the chance of injury etc. during handling.

In India, the use of anaesthesia in Fishery Science mainly for transportation dates back to the early 1960s. McFarland (1959, 1960) has contributed to the knowledge of lchthyo-anaesthetics. Bell (1967) reviewed the literature on the comparative effectiveness of different anaesthetics. Smith and Bell (1967) have discussed different properties of anaesthetics. Kewalramani and Gogate (1968) introduced a method of administration of anaesthetics directly into the body of rohu, catla and mrigal. The techniques of application of anaesthetics were described by Klontz and Smith (1968) and different stages of anaesthesia along with its hazards have been explained by McFarland and Klontz (1969). Bove (1962) and Huish (1972) suggested that effective concentration of anaesthetics used in pisciculture and research work depends on the size, maturity and the fish species.

Kewalramani and Gogate (1968) suggested that Novocaine is generally used in surgical operations and relief from pain; Barbiturates are usually given for alleviating mild discomforts, Phenobarbital and Barbital sodium are long-acting drugs; Pentobarbital and Thionyl induce anaesthesia for short or very short durations and Amobarbital is intermediate in action. Sreenivasan (1962) reported a dose for urethane ($100.0 mg.l^{-1}$), thiouracil ($10.0 mg.l^{-1}$), quinaldine ($5\text{-}10 mg.l^{-1}$) and hydroxyquinaldine ($1.0 mg.l^{-1}$) for sedation. Different anaesthetics (such as Novocaine, Amobarbital sodium and Barbital sodium) have been used by Kewalramani and Gogate (1968) to sedate certain Indian major carps. Schoettger and Julin (1967) studied the response of MS-222 (3 aminobenzoic acid ethyl ester methane sulphonate) in salmonids. The toxicity of MS-222 to some warm water and coldwater fishes was determined by Marking (1967). Locke (1969) suggested that quinaldine should replace MS-222 but fishes show responses in the former. Methyl pentynol is a useful anaesthetic in reducing activity in salmonids (Svendsen, 1969) but a higher dose of methyl pentynol as compared to MS-222 is required to produce similar sedative effects. Response levels of MS-222 were investigated by Sylvester (1975) in striped mullet (*Mugil cephalus*). Schoettger and Steucke (1970) suggested that methyl pentynol may be used along with MS-222. According to Laird and Oswald (1975), Benzocaine has ease of preparation, long storage life and safety. Meyer et al, (1976) reported that MS-222 is registered by The United States Environmental Protection Agency (USEPA) and The Food and Drug Administration (FDA) as a standard fish anaesthetic. But, Smit et al, (1977) have found that MS-222 causes changes in the quality of freshwater. Influence of water temperature, hardness and stocking density on the response of MS-222 in common carp (*Cyprinus carpio*), fathead minnow (*Pimephales promelas*) and rainbow trout (*Salmo gairdneri*) were worked out by Sylvester and Holland (1982). Effects of MS-222 and tagging on blood composition were investigated by Nieminen et al, (1982) in splake. Sharma et al, (1991) reported that MS-222 is effective in narcotizing *Cirrhinus reba* and *Barbus mahicola* at 50 ppm. In the view of Basavaruju et al, (1998) catla is more delicate than common carp and requires lower doses of Benzocaine for effective anaesthetization.

A comparative study on aquatic oxygen uptake under sedation in the early developmental stages of *Catla catla* (Ham) and *Labeo rohita* (Ham) has been made by Jha et al, (2004). Pandit and Ghosh (2005) also studied the effect of Benzocaine on opercular movement and oxygen consumption in *Channa punctatus* (Bloch). Kumari et al, (2009) investigated anaesthetic effect of Benzocaine on oxygen consumption of a carp, *Labeo bata*.

The anaesthetics change the physiological behaviour of fish but information on the effect of anaesthetics on the respiratory physiology of fish is limited (Wedemeyer, 1970; Houston et al, 1971). Smit, 1970, 1971; Parvathy et al, 2019). Jha and Ghosh (1993) and Ghosh and Jha (1994) have studied the effect of Benzocaine on the rate of oxygen uptake in the juveniles of *Labeo rohita* and *Catla catla* respectively. Literature regarding effect of anaesthetics on oxygen uptake in Indian air-breathing fishes is probably lacking. Pandit and Ghosh (1999) evaluated the effect of Benzocaine on the aquatic oxygen uptake in the juveniles of an air-breathing catfish, *Heteropneustes fossilis*.

Effect of Ammonia on Gills

Ammonia is the main excretory product of nitrogen metabolism in fishes, which is mainly synthesized in the liver and released through the gills. Ammonia is an important toxicant of water and it exerts its effect on the fishes. Fromm and Gilette (1968) suggested that ammonia can pass through the branchial membrane much easier than the ammonium ion due to its lipotropic property. In the opinion of Larmoyeux and Piper (1973), ammonia in the surrounding water of the fish hinders its depletion from the decomposition of protein by the fishes. So, the accumulation of nitrogen may induce alternations in branchial tissue and causes mortality of fish (Peters et al, 1984). Ammonia determines the toxicity level in the water and brings about problems in pisciculture by causing various diseases and altering the normal physiological activities of the fish. While acute exposure to ammonia causes proliferation of gill lamellae sublethal exposure reduces the oxygen-carrying capacity of haemoglobin; increases oxygen uptake, and heart-beat; urine output and excitability. The long term exposure of ammonia causes decreased growth, etc.

Brockway (1950) explained that ammonia may decrease 1/7th of the normal value of oxygen concentration of trout (*Salmo* sp.) blood and inhibit oxygen carrying-capacity of haemoglobin. Lloyd (1961) observed that reduced environmental oxygen concentration enhances the toxicity of ammonia and suggested that ammonia primarily acts on the respiratory system. The effect of long term exposure to ammonia has been investigated by Burrows (1964) on Chinook salmon (*Oncorhynchus tshawytscha*), Reichenback-Klinke (1967) on different fishes, Flis (1967, 1968a and 1968b) on common carp (*Cyprinus carpio*), Bullock (1972) on salmonids and Larmoyeux and Piper (1972) on rainbow trout (*Salmo gairdneri*) and reported changes in gill structure. Investigations of Fromm and Gillette (1968), Wilson et al, (1969) and Santhi et al, (1992) indicated that ammonia may act on the nervous system of fishes. Lloyd and Orr (1969) observed that rainbow trout (*Salmo gairdneri*) exhibited an increased permeability to water under exposure to ammonia and suggested that if such condition exceeds urine output, the fish might die. Histopathological changes in the gills due to prolonged exposure to ammonia was observed by Hughes (1972), which may reduce the oxygen diffusing capacity of gills, leading to insufficient oxygen uptake. Meyer and Kramer (1973) found that long-term exposure of rainbow trout to a combination of elevated ammonia and reduced oxygen concentration results in a lowering of active metabolism. Hughes and Perry (1976) measured the diffusing capacity of the gills of trout under long-term exposure to ammonia and reported damage to the gills.

The median lethal threshold of unionized ammonia for rainbow trout (*Salmo gairdneri*) was found to be 0.45mg.l^{-1} by Smart (1975). Robinette (1976) observed an inhibitory effect of sublethal ammonia (120µg.l^{-1}) on the growth of channel catfish (*Ictalurus punctatus*). The growth and metabolism of juvenile rainbow trout (*Salmo gairdneri*) under sublethal ammonia (110 µg.l^{-1}) was studied by Schulze-Wiehenbrauk (1976). Gill structure of rainbow trout (*Salmo gairdneri*) has been studied by Smart (1976) in which he noticed minor gill damages at an acute lethal concentration ($0.43\mu g.l^{-1}$) of ammonia followed by severe histopathological changes and impaired oxygen uptake along with the high incidence of diseases in its long-term exposure (0.25 to $0.30\mu g.l^{-1}$). Prolonged exposure of ammonia showing alterations in physiological processes has been observed by Burkhalter and Kaya (1977) on fertilized eggs and sac fry of rainbow trout (*Salmo gairdneri*). Colt and Tchobunoglous (1978) studied histopathological changes in viscera during the growth and survival of channel catfish (*Ictalurus punctatus*). Anderson (1979) studied the growth of Dover sole (*Solea solea*) and turbot (*Scophthalmus maximus*) under the effect of ammonia. He explained that long-term exposure to ammonia decreases the rate of growth. The toxicity of ammonia on fingerlings of Coho salmon (*Oncorhynchus kisutch*) was worked out by Buckley (1978). Thurston and Russo (1978) reported gill hyperplasia, variation in respiratory surface area and change in the gill epithelium due to acute toxicity of ammonia and nitrite to the fries of cutthroat trout (*Salmo clarki*). Respiratory alteration with gill damage was considered to be the first symptom by Maki (1979) in fish underexposure of lethal and sublethal concentrations of ammonia. The toxic effect of ammonia with pH and calcium has been worked out by Tomasso et al, (1980) on channel catfish (*Ictalurus punctatusus*). Thurston et al, (1981) studied the effects of pH on the toxicity of ammonia in fishes and explained that pH is a major factor that affects the toxicity of ammonia. Thurston et al, (1981) reported increased toxicity of ammonia to rainbow trout (*Salmo gairdneri*) resulting from a lowered amount of dissolved oxygen. Long-term exposure of ammonia on the developmental stages of rainbow trout (*Salmo gairdneri*) has been worked out by Calamari et al, (1981). Effect of acute toxicity of ammonia to rainbow trout (*Salmo gairdneri*) and fathead minnow (*Pimephales promelas*) was worked out by Thurston and Russo (1983) and Thurston et al, (1983) respectively. The effect of prolonged exposure to ammonia has been worked out by Thurston et al, (1984) on rainbow trout (*Salmo gairdneri*); Thurston et al, (1986) on fathead minnow (*Pimephales promelas*) and Wajsbrot et al, (1993) on gilthead seabream (*Sparus aurata*). Fish mortality due to acute exposure to ammonia on goldfish (*Carasius auratus*) has been observed by Tarazona et al, (1987). Acute toxicity of ammonia was investigated by Ostrensky and Brugger (1992) on the viability of cultivation of silverside (*Odontesthes argentinensis*) to evaluate the criterion for aquaculture.

In India, Ram and Sathyanesan (1986, 1987a and 1987b) have studied the effect of ammonium sulphate on nuclear changes in oocyte; testicular development of follicles and histopathological changes in liver and thyroid of *Channa punctatus*. The toxicity of ammonium sulphate on histopathological changes in the epidermis and inner epithelium of air sac of *Heteropneustes fossilis* was observed by Banerjee and Paul (1993) and Paul and Banerjee (1995) respectively. Reddy et al, (1996) have studied the effect of ammonia exposure on the carbohydrate metabolism of *Cyprimus carpio*.

Packing Conditions for Transportation of Live Fish

Transportation of different life stages of fish is a problem due to high mortality despite the care exercised during the process. Nearly all the transportation methods involve the use of some device for aeration to ensure an adequate oxygen supply for the fish. Some chemicals or anaesthetics are used for anaesthetization of fishes to lower their activity. It seems that safe transport of fishes suffers from many integrated factors such as size or weight of the fish, size of the container, distance to be transported, water temperature, pH, the supply of oxygen, accumulation of waste products and fatigue from continued splashing etc.

The fishes are conditioned before transportation to adjust them in a limited area. The process of conditioning has been reported by Jagannadhan (1947), Saha and Chowdhury (1956), Alikunhi (1957), Hora and Pillay (1962) and Srivastava and Karamchandani (1964) Barton (1997) etc. The conditioning depends upon the size and health of spawn, fry, fingerlings and brooders etc. Jagannadhan (1947) suggested that conditioning of about 48 to 72 hours is necessary before transportation while 6 hours conditioning has been suggested by Alikunhi (1957).

Handling, an important and essential process during transportation generally increases the metabolic rate of fishes. Black (1958) explained that during handling, many fishes accumulate a dangerous level of lactic acid in their blood. But, combined methods of drugs and aeration increase the number and weight of fishes during transportation.

Anaesthetics are used to reduce the metabolic activity of fishes during transportation. Urethane lowers the oxygen uptake of fishes and may be used during the transportation of fishes under hot weather. Sodium amytal and sodium seconal have been tested by the California Department of Fish and Game to transport trout (*Salmo gairdneri*). Osborn (1951) used 388 ppm of thiouracil on stone rollers and big mouth shiners by which oxygen uptake of these fishes get reduced by about 20% and the fishes probably develop an ability to withstand a lower concentration of dissolved oxygen. Natarajan and Ranganathan (1960) explained that quinaldine (a fully safe tranquilizer of coaltar product) can be effectively and economically used in the transportation of live fishes of all the stages because it reduces activity and mortality. A method of anaesthetization of carps was introduced by Kewalramani and Gogate (1968) using Amobarbital sodium, Barbital sodium and Novocaine.

Various types of appliances are used for transportation in different parts of India and abroad. In the USA and Canada, cuvettes of Styrofoam are used for transporting salmon eggs. In Philippines and Indonesia, boats are designed to allow free water circulation in their holes where fishes are held. Hora and Pillay (1962) reported that tar-coated, watertight plaited bamboo baskets of 15litre capacity are used in Indonesia. In China, wooden tubs (2m height and 2m diameter) and bamboo baskets are commonly used. Konradt (1968) reported that carp larvae ($3000 \pm 1000.l^{-1}$) are transported in plastic bags having 1:1 ratio of water and oxygen in the USSR.

Earthen vessels (hundies) are extensively used in West Bengal to transport spawn, fry and fingerlings. State Fisheries Department of Orissa improvised an open transport motor van with a small semi-rotatory pump. Department of Fisheries, Maharastra introduced container of economical polythene bags to transport major carps. Alkathene bags (74.0x46.0cm) are now widely used in India. Open tin containers of various designs were used by the Madras Fisheries Department. To transport fingerlings and adults, open canvas containers (1x1x1.25m) and galvanized iron drums (180 litre capacity) are used in Punjab and Madhya Pradesh. The transport of spawn and fry in oxygen sealed packs gained popularity in India. Khan (1939 and 1946) suggested oxygenated cylindrical water containers (45.72x35.36cm) for transporting fish seed. Ganapati and Chacko (1951) made a comparative study of the transport of fish seed in the ordinary tin container with or without oxygen. CIFRI introduced the use of kerosene tins filled with airtight screw-capped lids for transportation. Saha and Chowdhury (1956) described traditional methods of transporting fry and fingerlings in hundies. Saha and Sen (1958) described three types of materials (glass carboys, latex rubber and plastic bags) for the transport of spawn and fry under oxygen pressure.

Various types of packing methods have been applied in many countries and put to successful practice test. Long ago, McDonald (1884) demonstrated that carp would survive a long duration of time in the limited amount of water while preparing a report on carp. Mulligan and Majid (1936) used glass containers to transport *Gambusia*. A description of typical transport method and equipment for *Salmon* eggs was given by Burrows (1949). Fertilized eggs are usually transported in iso-thermic conditions in which a humid atmosphere can be maintained (Martyshev, 1983). Mitra (1942) transported fingerlings in a suitably designed oxygenated container. Sixty fingerlings of *Cprinus carpio* were transported from Ooty to Kumaon by Raj and Cornelius (1949) in sealed containers with no mortality. Vass (1951) reported reduced mortality of fish during a journey of 12 to 50 hours under sealed jerry cans.

Lindstedt (1914), suggested a formula ($R = \frac{V}{N}, \frac{V}{N} \alpha W^{\frac{2}{3}}$ where R= rate of respiration, V= oxygen entering h^{-1} into a volume of water, N=number of fry and W= weight of fry) for the transport of a consignment of fry. Basu (1950), Albaster and Herbert (1954) and Saha et al, (1956) studied the effect of carbon dioxide on oxygen uptake to the transportation of fishes. Vass (1951) tried to establish the factors governing the number and weight of fry, the period of transport and the pressure of oxygen. He also suggested that the addition of secondary sodium phosphate or disodium hydrogen phosphate (Na_2HPO_4) in water increases the duration of transport due to its buffering action. Khan (1940), Saha et al, (1956), and Singh (1977) worked out on the oxygen uptake of fishes to study the effect of dissolved oxygen on transportation. Jagannadhan (1947) reported that some air space should be left above the water surface in packed containers for safe transport. Saha and Chowdhury (1956) discussed that ingredients present in the supply of water are essential for the spawn during transportation as instantaneous mortality of spawn in distilled water was observed. They showed that pulverised soil increases the survival of spawn but acidic pH causes their instantaneous mortality. Saha et al, (1956) demonstrated that pulverised earth, activated charcoal and Amber lit tend to absorb carbon dioxide and ammonia from the medium and thus increases the survival period of fish during transportation. Basu (1951) concluded that red soil coagulates the suspended organic matter and thus keeps down pollution. Saha et al, (1956) summarized that the number of fries to be packed is more or less proportional to the critically given oxygen pressure. Ramachandran (1969) observed the total mortality of fish in the pressure of ammoniacal nitrogen (over $20 mg.l^{-1}$) though there was sufficient oxygen in the container of an experimental packed consignment.

In India, Central Inland Fisheries Research Station organized a fish seed syndicate at Calcutta in 1952 to supply fry and fingerlings of Indian major carps to deficit states for piscicultural purposes. Air transport of carp fry in packed plastic bags for the export of live tropical fish to England and other countries was put into practice by the Department of Fisheries, Bombay way back in 1952. Ranade and Kewalramani (1956) summarized that the air transport of carp fry is both simpler and cheaper than the rail transport method. According to Martyshev (1983), live fishes are also transported without water in different containers by covering the bottom with wetting materials. Privol'nev (1964) recommended that live fishes may be transported in humid conditions. Murrels are transported in pots with some water or in the basket with wet green weeds (Satyanarayana, 1996). Kulkarni and Ogale (1979) made air transport of mahaseer in India. An extensive study has been made to pack fingerlings of *Labeo rohita* and *Catla catla* by using anaesthetic and commercial oxygen for transportation of different hours (FTR OF ICAR, 1994). Optimization of packing conditions for live transportation of *Anabas testudineus* and *Channa punctatus* has been made by Pandit and Ghosh (2012).

In the present book, an attempt was taken to evaluate the oxygen requirement of two economically important air-breathing fishes namely, *Anabas testudineus* and *Channa punctatus* particularly and other air-breathing fishes in general to body weight and group size during their early developmental stages. The effect of anaesthetic on the oxygen uptake and the effect of ammonia on the structure and function of gills in *Channa punctatus* will be investigated.

Finally, these findings will be utilized to evaluate suitable packing methods for the transportation of *Anabas testudineus* and *Channa punctatus*. The information of this work may be applied for live fish transportation of all the fishes.

3

Oxygen Uptake in Relation to Body Weight

Introduction

Oxygen uptake of fish(s) displays its basal metabolic status and is taken into consideration as one of the signs of the overall fitness or/and well-being of the fish. It can also be beneficial to evaluate the physiological state of a living organism, facilitates in comparing the susceptibility or resistance potentiality and additionally beneficial to correlate the behaviour of the animal, which ultimately serve as predictors of functional disruptions of population. Analysis of Oxygen uptake machine to assess the primary harm or demand inflicted at the animal that can both boom or lower the oxygen uptake (Neelima et al, 2016).

The maximum vital single issue in transporting fish is imparting an adequate level of dissolved oxygen. Heavier fish transported in warmer water want more oxygen. For instance, if the water will increase by temperature 10°C (e.g., from 10 to 20°C), Oxygen uptake is roughly doubled.

Respiration of air-breathing fishes has fascinated researchers for a minimum of a century; the earliest attempts to measure oxygen uptake ($\dot{V}O_2$) and respiratory partitioning date to the 19th century (Baumert, 1853). The available general respirometry methods have been explained and reviewed by numerous authors (Cech, 1990; Lighton, 2008; Lefevre et al, 2016; Zhang et al, 2019), and researches of the respiration of air-breathing fishes had been reviewed in element within side the iconic book by Graham (1997). Air-breathing fish are generally characterized by a significant reduction in gill surface area with a purpose to keep away loss of oxygen taken up in the air-breathing organ (ABO) at the site of the gills in hypoxic water (Scott et al, 2017; Pelster et al, 2020).

It has always been a problem for fish culturists to prevent high mortality of early developmental stages during transportation. Fishes die due to hypoxia or hypercarbic conditions of the water. There exist oxygen and carbon dioxide optima for different fish species (Wells, 1913). A fish mainly utilizes oxygen from water and thus decrease of oxygen below a certain level leads to unfavourable conditions for the fish and may result in their mortality. The metabolic rate of early developmental stages varies in response to ontogenetic processes and abiotic components. The early developmental stages include changes in their morphology and behaviour as well as associated changes in physiology and metabolism. The resistance to unfavourable conditions varies with individual species, weight and size. Smaller fishes are more resistant in per unit body weight than the larger ones.

It was demonstrated from the work of Rübner (1883) that the oxygen uptake of a species shows dependency on body weight. He observed metabolic rate/body weight relationship to be $W^{0.67}$ in different breeds of dogs. Zeuthen (1953) estimated oxygen uptake to body weight and reported slopes (=b) of 0.70, 0.75 and 0.95 for the animals of body weight up to 1 µg, above 40 mg and 1 to 40 mg respectively. He concluded that the metabolic rate of animals decreases with increasing body weight. Winberg (1956) noted a relative slope value of 0.81 for several fish groups. Hemingsen (1960) calculated a slope of 0.75, probably common to all the animals and suggested that weight-specific metabolic rates or metabolic rate per unit body weight ($mlO_2.g^{-1}.h^{-1}$) decrease with increasing body weight. The finding supports the statement of Prigogine and Wiame (1946), that for some thermodynamic reasons, the weight-dependent metabolic rate never increases with increasing body weight. Moss and Scott (1961) and Beamish (1964a) suggested that metabolism to body weight in fishes is specific for different species. Beamish (1964a) also observed that different species of fish from the same habitat have different values of regression coefficient or slope (=b) for oxygen uptake and these differences are probably due to their genetic variation.

The oxygen uptake of a fish may be differentiated as the standard metabolic rate (SMR) or minimum metabolism, routine metabolic rate (RMR) or metabolism of normal activity (MNA) and active metabolic rate (AMR) or metabolism of maximum sustained activity (MMSA). SMR is related to the metabolism of inactive fish with minimum metabolism to sustain their survival. RMR is the metabolic rate for normal swimming, normal-feeding and spontaneous activities etc. AMR represents metabolic rate under sustained and enforced situations (Fry, 1971). According to Bushell, et al, (1994), the maximum metabolic rate (MMR) is the active metabolic rate (AMR) for a limited period. The metabolic rate of animals depends on habitat temperature. Any increase or decrease in temperature induces a considerable change in the metabolic rate.

Oxygen uptake of spawn, fry and fingerlings of carps have been studied by many investigators (Basu, 1951; Hora and Pillay, 1962; Singh, 1977; Bhaumik et al, 1986). Generally, fry, fingerlings, juveniles and table fishes are typically transported live in different types of traditional containers by sprinkling and circulating water within the container (Jhingaran, 1975; Jhingaran and Pullin, 1985; Ou and Li, 1996). There have been few attempts for the transport of live fishes by various methods, particularly brood fishes with some degree of success (Mammen, 1962; Ramchandran, 1972).

Attempts have been made to determine oxygen uptake in some Indian air-breathing fishes namely, *Anabas testudineus* (Hughes and Singh, 1970b), *Heteropneustes fossilis* (Hughes and Singh, 1971) and *Clarias batrachus* (Singh and Hughes, 1971) etc. under different experimental conditions. The relationship between oxygen uptake and body weight in different air-breathing fishes has been observed by various workers. Munshi and Dube (1973) reported a slope of 0.67 for aquatic oxygen uptake in running water with access to air and a slope of 0.53 without access to air in *Anabas testudineus*. Under similar conditions, two slope values of 0.869 and 0.841 were reported to show the relationship of oxygen consumption and body weight in magur (*Clarias batrachus*)(Munshi et al, 1976). Ojha et al, (1977) also obtained 0.671 and 0.595 respectively as slope values for body weight and aquatic oxygen uptake relationship in *Colisa fasciatus* under similar conditions.

Munshi et al, (1978) observed that aquatic oxygen uptake in *Heteropneustes fossilis* increases by 1.084 and 0.986 in the juveniles and adults respectively when access to air was allowed but the values were comparatively less (0.478) in juveniles than in adults (0.799) when air-breathing was prevented. In *Channa marulius*, the total oxygen uptake is much higher in their juveniles than that of the adults (Ojha et al, 1979). While working with *Channa gachua*, Ojha et al, (1978) also calculated two slope values (0.681 and 0.641) under access to air and its prevented conditions respectively. The relationship between aquatic oxygen uptake and body weight in *Channa punctatus* and *Pseudapocryptes lanceolatus* has been determined by Hakim et al, (1983) and Ojha and Haque (1994) respectively.

Bimodal oxygen uptake to body weight has been determined by various workers in different air-breathing fishes (Lamholt and Johansen, 1976; Ojha et al, 1978; Biswas et al, 1979; Ghosh et al, 1980; Munshi et al, 1982; Patra et al, 1983; Ghosh and Munshi, 1987 and Ojha et al, 1988). Bimodal oxygen consumption concerning bodyweight in *Channa punctatus* at two seasonal temperatures has been investigated by Ghosh et al, (1990). Ghosh and Munshi (1987) made a study on the bimodal oxygen uptake of *Anabas testudineus* to body weight and obtained two slope values (0.933 at 21.0 ± 1.0°C and 0.899 at 31.5 ± 1.0°C) at two seasonal temperatures. Bimodal oxygen uptake to body weight was estimated by Ojha et al, (1988) in *Colisa fasciatus* and found that with the unit increase in body weight, aquatic, aerial and total oxygen uptake increased by powers of 0.537, 0.266 and 0.372 respectively.

Besides the above Indian air-breathing fishes, the oxygen uptake of early developmental stages (body weight: 0.7 to 5.0g) of *Periophthalamus sobrinus* has been studied by Gordan et al, (1969). Aerial and aquatic oxygen uptake of different species of *Periophthalamus* was observed by Tamura et al, (1976) and Gordan et al, (1978). Bimodal gaseous exchange was investigated by Burggren (1979) to environmental oxygen and carbon dioxide in *Trichogaster trichopterus*. Houlihan and Morrison (1982) observed aerial and aquatic respiration in some littoral fishes to their vertical zonation. Aerial and aquatic oxygen uptake of *Lipophrys pholis* has been estimated by Laming et al, (1982) and Pelster et al, (1988). The oxygen uptake of *Periophthalamus cantonensis* was studied by Chan (1990). The gill respiratory parameters during progressive hypoxia in a facultative air-breathing fish, *Hypostomus regani* was studied by Mottias et al, (1998). However, in the above-mentioned works, the relationship between oxygen uptake and body weight has not been established.

Basu (1950) determined the oxygen uptake in *Catla catla, Cirrhina mrigala, Labeo bata* and *Labeo rohita*. The oxygen uptake in the fingerlings of different Indian major carps has been determined by Motwani and Bose (1957), Saha (1966) and Singh (1977) etc. Roy and Munshi (1984) reported two slope values in *Cirrhinus mrigala* during summer (b=0.8113) and winter (b=0.7961) with an increase in body weight from 5.0 to 180.0 g. Pandey (1988) calculated a slope of 0.5976 for early developmental stages (5.0 to 260.0 mg.) and 0.8576 for mature fishes (16.0 to 425.0 g) of *Labeo rohita*. Kunwar et al, (1989) reported slope values of 0.7469 and 0.8515 respectively in *Catla catla* (6.5 to 240.0 g) and *Labeo rohita* (16.0 to 425.0 g). Jha (1995) obtained comparatively fewer slope values in *Catla catla* (b=0.5612) and *Labeo rohita* (b=0.5384) in the weight range of 15.0 to 1000.0mg and 5.0 to 903.0 mg respectively.

Information on the relation between oxygen uptake and early developmental stages in air-breathing fishes is limited (Gordan et al, 1969; Chan, 1990 and Rombough-Peter, 1998). Therefore, the objective of this work is to determine oxygen uptake in the early developmental stages of two economically important air-breathing fishes, *Anabas testudineus* and *Channa punctatus*.

The information will estimate the amount of oxygen requirement of these fishes for their safe transport for longer periods. It will also be helpful in the storage and breeding of fishes at a commercial scale.

Plate – I

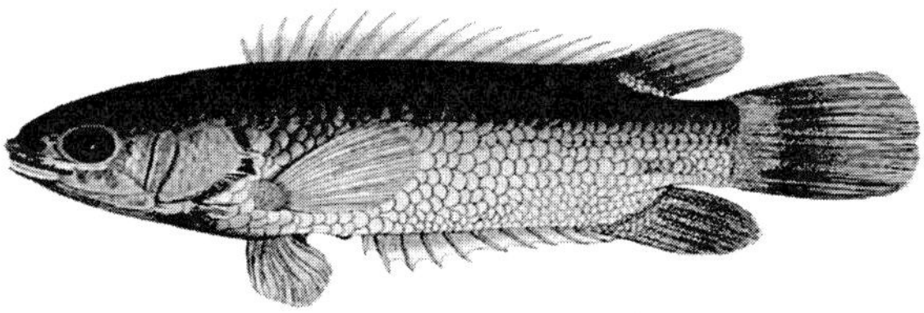

Figure 1: *Ananbas testudineus* (Bloch)

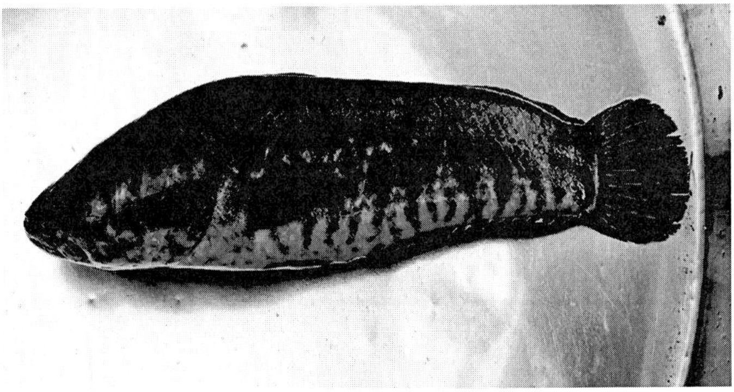

Figure 2: *Channa punctatus* (Bloch)

Materials and Methods

To obtain the early developmental stages of *Anabas testudineus* (Bloch) (Plate-I, Figure 1) and *Channa punctatus* (Bloch) or *Channa punctatus* (Menon, 1974), (Plate-I, Figure 2) the following method of induced breeding was applied. Proper sanitization of all the equipment and accessories is one of the essential steps for the determination of oxygen uptake of fishes in this pandemic of COVID-19.

Induced breeding of air-breathing fishes: Mature *Anabas testudineus* and *Channa punctatus* were collected from rural areas of Katihar district of North Bihar. These were acclimated in large plastic pools. The fishes were fed with the earthworm, chopped trash fish and liver of goat. After acclimation for a fortnight, sex-wise segregated fishes were stocked in separate plastic pools. The mature females were identified by their soft and swollen abdomen while mature males oozed milt on gentle pressure.

The *Anabas testudineus* were bred using the hypophysation method (Moitra et al, 1979). For this, the weight and the sex-wise ratio of male and female were kept in 1:2. The males were injected with a single dose of 75-80 $mg.kg^{-1}$ body weight of heteroplastic pituitary extract while in females, the required dose (75-80 $mg.kg^{-1}$) was given in two installments at an interval of 4 to 5 hours. Spawning occurred after 6-10 hours of injection. The parents were lifted from the plastic pools after spawning, as they devour their young ones, Plankton were added to the plastic pool after the disappearance of the yolk sac.

In *Channa punctatus*, a dose of 80-100 $mg.kg^{-1}$ body weight of heteroplastic pituitary was administered in females in two installments but the males were injected with a single dose of 80-100 $mg.kg^{-1}$ body weight. When 3-4 sets of such mature males and females were kept in a ratio of 1:1 in a cemented tank (250×150×100 cm) having stagnant water, they bred successfully. The tank receives runoff water from a tube well and had a rich algal bloom. The hatchlings of *Channa punctatus* were left in their natural environment with their parents, as they showed parental care till they attain a total length of about 30 mm and an average bodyweight of 230 mg.

The different stages of Indian major carps were classified based on their size. According to Chakraborty and Murty (1972), spawn or hatchlings, early fry, fry, advanced fry and fingerlings measure between 5.0 to 7.5, 8.0 to 13.5, 14.0 to 25.0, 25.5 to 40.0 and 41.0 to 100.0 mm respectively. However, such a scheme is not used in *Anabas testudineus* and *Channa punctatus* due to the smaller size of their parents. In this experiment, early fries have been taken and the term early developmental stages or juveniles have been used.

Following two experimental set-ups were made for measuring oxygen uptake in different weight groups of early developmental stages of *Anabas testudineus* and *Channa punctatus*.

A. Measurement of aquatic oxygen uptake: A cylindrical glass respirometer with the continuous water flow system (Munshi and Dube, 1973) and as modified by Jha and Ghosh (1991) (Figure 1) was used to measure the aquatic oxygen uptake of the hatchlings of *Anabas testudineus* and *Channa punctatus*. The early fries were taken in the respirometer and were filled with water. The respirometer was connected to a water reservoir or aquarium (60×30×30 cm) kept at a height, maintaining a constant hydrostatic pressure.

Both the ends of the respirometer were connected with rubber tubes. Three sets of the conical flask with a delivery tube system were also connected with a rubber tube in a specific manner to collect samples of inspired and expired water. A black cloth was kept from three sides of the respirometer, so that the fish do not feel disturbed, leaving one side open for observation. The rate of water flow through the respirometer was adjusted according to the size and the number of early fries.

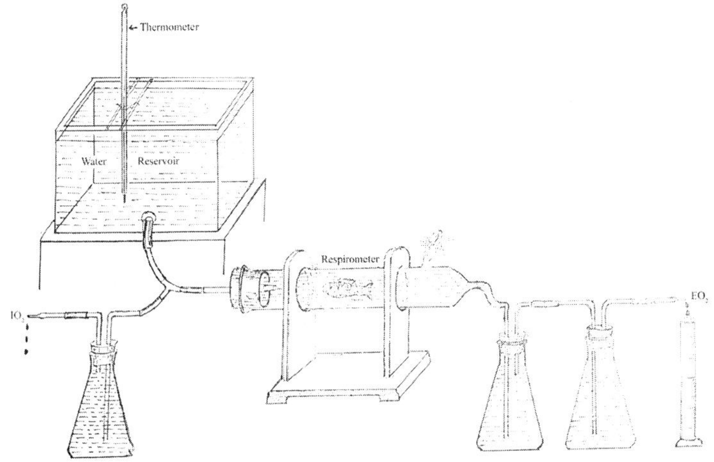

Fiure 1: An experimental set-up for measuring aquatic oxygen uptake through a cylindrical glass respirometer with continously water flowing system

B. Measurement of bimodal oxygen uptake: Simultaneous oxygen uptake from the air and still water in early fries of *Anabas testudineus* and *Channa punctatus* were determined in a closed glass respirometer (Ghosh and Munshi, 1987) of 3 litre capacity. The early fries were kept in the respirometer containing 2 liters of tap water. A Styrofoam or thermocol float with a peripheral semicircular hole was placed over water to separate the air-water interphase of the respirometer. The air phase of the respirometer was connected to a differential fluid (kerosene oil) manometer using a rubber tube. A 'T' was placed between the manometer and outlet from the respirometer. Another tube is attached to the 'T', the end of this tube is fitted with a plastic stopper. The fries were allowed to take air gulps through the semicircular hole. Pellets of potassium hydroxide (KOH), were kept over the float in a Petri dish, which absorbs the carbon dioxide (CO_2) liberated during respiration. The respirometer was placed in a constant temperature water bath. An outlet from the bottom of the respirometer was attached to a series of conical flasks to collect water samples (Figure-2). The aerial oxygen uptake was measured by determining the amount of pure oxygen needed to neutralize the rise in the level of fluid in the manometer and by the use of the combined gas law equation and vapour pressure (Dejours, 1975).

The amount of dissolved oxygen in both the methods was determined in inspired and expired water by Winkler's volumctric method (Welch, 1948).

The rate of oxygen uptake to individual fish and body weight was calculated from the difference in dissolved oxygen content between the inspired and expired water.

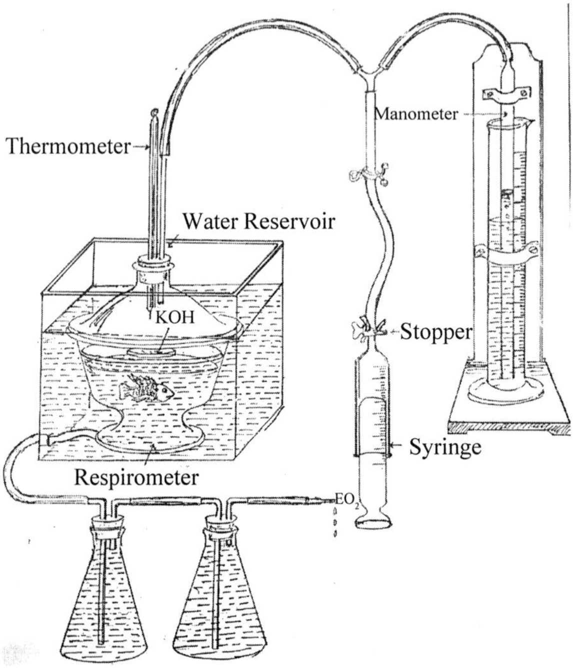

Figure 2: An experimental set-up measuring biomodal (simultaneously from air and still water) oxygen uptake through a closed glass respirometer

All the experiments were conducted between 9 to 11 a.m. (i. e. fixed hours of the day) in a temperature-controlled room with lights on. Early fries of *Anabas testudineus* and *Channa punctatus* ranging between 4.0 to 1000.0 mg in body weight and total length between 4.0 to 45.0 mm were used for this work. The early fries were always taken in groups (5-40) to achieve better results as the determination of oxygen uptake of such a small specimen was difficult (Kamler, 1976; Jha, 1995). After experimentation, the early fries were removed from the respirometer. Only living early fries were fixed in 5% formaldehyde, dried on a blotting paper and weighed up to the nearest mg. The dead specimens (early fries) were discarded. All the experiments were conducted using fresh tap water (pH=7.2 ±0.08, DO = 7.85 ±0.48 ppm. Free CO_2 = 2.4±1.4 ppm).

Analysis of data: Mean values of a series of observations made on the early developmental stages of different body weights were calculated. Data of oxygen uptake ($\dot{V}O_2$) were expressed in terms of per unit time (μlO_2.fis $h^{-1}.h^{-1}$)

and per unit body weight ($\mu lO_2 . mg^{-1} . h^{-1}$). The rate of oxygen uptake and body weight was transformed into the log to stabilize the variance before statistical analyses. The relationship between oxygen uptake and body weight may be expressed by the following allometric formula:

$Y = {}_aX^b$ (Wohlschlag and Juliano, 1959)

Or, $\dot{V}O_2 = {}_aW^b$

Or, Log $\dot{V}\,O_2$ = log a + bx log W

Where,

Y ($\dot{V}\,O_2$) = Metabolic magnitude(oxygen uptake. $24h^{-1}$ individual^{-1} or oxygen uptake. h^{-1}).

a = A constant of the ratio i. e., intercept or rate of oxygen uptake for unit body weight.

b = Regression coefficient, slope or weight exponent. It has a characteristic under various physiological states or functions of different environmental factors.

X (=W) = Bodyweight.

Other workers have expressed the allometric growth relationship in the following ways:

$M = K\,W^n$ (After Hoar, 1966)

$Q = a\,W^K$ (Belokopitin, 1968)

Where, M or Q= Metabolic magnitude

n or k = Regression coefficient between metabolism and body weight.

K or a = Constant of a ratio for oxygen uptake and body weight.

In the metabolism/ bodyweight dependence relationship weight (W) stands for the variable, while coefficient (b) and intercept (a) are known values recorded in the metabolism/ bodyweight ratio.

The multiple regression analysis was made on Graphpad Prism 5.04 software to establish the relation between oxygen uptake and body weight.

Observations and Results

The data on oxygen uptake to body weight in *Anabas testudineus* and *Channa punctatus* have been summarized in Tables 1 to 6 and their respective computed values are compiled in Tables 7 and 9. The data of observed values indicate that the rate of oxygen uptake per unit time ($\mu lO_2.mg^{-2}.h^{-1}$) increases with an increase in body weight while the weight-specific oxygen uptake ($\mu lO_2.fish^{-2}.h^{-1}$) generally decreases with increasing body weight. The estimated values of oxygen uptake for a fish of 1.0, 10.0, 100.0 and 1000.0 mg body weights are shown in Table–10.

A. Oxygen uptake in *Anabas testudineus*

a. Aquatic oxygen uptake from continuously flowing water at 28.0±1.0°C

Oxygen uptake per unit time ($\mu lO_2.fish^{-1}.h^{-1}$): It ranged from 25.216±6.711 to 251.444±9.655$\mu lO_2.fish^{-1}.h^{-1}$ with an increase in body weight from 9.56±4.88 to 834.83 ± 5.30 mg. The average oxygen uptake was found 143.220±64.403 $\mu lO_2.\ fish^{-1}.h^{-1}$ for a fish with an average body weight of 320.11±268.87 mg (Table – 1). The log/ log plot of aquatic oxygen uptake versus body weight gave a straight line when the different scores were fitted by the least square regression method. The slope (=b) of the regression line was 0.4522 and the estimated value for the rate of aquatic oxygen uptake for a fish of 1.0, 10.0, 100.0 and 1000.0 mg body weight was 11.907, 33.731, 95.554 and 270.688 $\mu lO_2.\ fish^{-1}.\ h^{-1}$ respectively (Table – 10). There exist a high degree of correlation between body weight and the rate of oxygen uptake ($r=0.9741,\ p<0.001$) (Table – 7). The relationship between the two variables can be expressed as:

$$\dot{V}O_2 = 11.9070W^{0.4522}$$

$$\text{Or, Log } \dot{V}O_2 = 1.0758 + 0.4522 \times \log W$$

$$0.3827 < b < 0.5217$$

The above equation indicates that for a unit increase in body weight, the rate of aquatic oxygen uptake through gills increased by a power of 0.4522.

Oxygen uptake per unit body weight ($\mu lO_2.\ mg^{-1}.\ h^{-1}$): The weight-specific oxygen uptake decreases from 3.087 ± 1.132 to 0.301 ± 0.009 $\mu lO_2.mg^{-1}.h^{-1}$ with an increase in body weight from 9.56 ± 4.88 to 834.82 ± 5.30 mg. On average, a fish of 320.11 ± 268.87 mg body weight consumes about 0.952 ± 0.801 $\mu lO_2.mg.h$ oxygen (Table – 1). The log/ log plot of oxygen uptake per unit bodyweight gave a straight line with a slope of -0.5726 (Table – 7). The two variables show negative and highly significant correlation ($r = 0.9859,\ p < 0.001$) and may be represented as:

$$\dot{V}O_2 = 13.6575\ W^{-0.5726}$$

$$\text{Or, Log } \dot{V}O_2 = 1.1354 - 0.5726 \times \log W$$

$$-0.6370 < b < -5082$$

Table 1: Rate of aquatic oxygen uptake ($\mu lO_2.fish^{-1}.h^{-1}$ and $\mu l\ O_2.mg^{-1}.h^{-1}$) in relation to body weight during early developmental stages of *Anabas testudinus* at 28.0±1.0°C.

S. No.	Body wt. of a fish (mg)	AQUATIC OXYGEN UPTAKE						
		$\mu lO_2.fish^{-1}h$			$\mu lO_2.mg^{-1}.h^{-1}$			$\mu lO_2.g^{-1}.h^{-1}$ Estimated
		Calculated	Estimated	95% CL for estimated value	Calculated	Estimated	95% CL for estimated value	
1	9.56±4.88	25.216±06.711	33.0434	22.2764 49.0145	3.087±1.132	3.7508	2.6025 5.4059	2639.037
2	30.38±6.88	60.698±11.496	55.7479	38.9132 79.8659	2.024±0.073	1.9344	1.3861 2.6995	1998.288
3	56.58±6.63	88.279±3.359	73.8594	52.1986 104.5086	1.559±0.207	1.3547	0.9829 1.8689	1560.168
4	85.03±10.35	94.832±3.359	88.7944	63.0928 124.9659	1.053±0.214	1.0730	0.7817 1.4729	1115.343
5	126.83±11.00	121.671±4.362	106.3972	75.8328 149.2806	0.966±0.059	0.8534	0.6235 1.1681	959.301
6	164.86±11.25	137.271±3.052	119.7911	85.4461 167.9411	0.895±0.047	0.7344	0.5369 1.0045	832.677
7	223.05±5.73	151.947±0.914	137.3414	97.9338 192.6061	0.0695±0.015	0.6177	0.4515 0.8451	690.191
8	329.00±2.50	132.616±3.041	163.7321	116.4825 230.1476	0.403±0.006	0.4945	0.3606 0.6780	403.088
9	465.17±4.25	165.285±2.965	191.4934	135.7108 270.2045	0.355±0.003	0.4055	0.2947 0.5580	355.324
10	505.67±4.5	186.241±3.390	198.8609	140.7675 280.9285	0.360±0.016	0.3866	0.2806 0.5325	368.308

S. No.	Body wt. of a fish (mg)	AQUATIC OXYGEN UPTAKE						
		$\mu lO_2.fish^{-1}h$			$\mu lO_2.mg^{-1}.h^{-1}$			$\mu lO_2.g^{-1}.h^{-1}$
		Calculated	Estimated	95% CL for estimated value	Calculated	Estimated	95% CL for estimated value	Estimated
11	614.83±3.75	205.152±5.450	217.2400	153.3015 307.8452	0.314±0.007	0.3457	0.2502 0.4775	333.671
12	712.67±6.03	239.204±5.035	232.6834	163.4507 329.9845	0.336±0.064	0.3176	0.2294 0.4399	335.646
13	834.83±5.30	251.444±9.655	249.4675	175.0183 355.5854	0.301±0.009	0.2901	0.2089 0.4030	301.191
Average	320.11±268.87	143.220±64.403	143.727±70.780		0.952±0.801	0.966±0.965		914.692±740.780

Therefore, a unit increase in body weight results in a decrease in weight-specific oxygen uptake by a power of -0.5726. The weight-specific oxygen uptake capacity of 1.0, 10.0, 100.0 and 1000.0 mg fish was calculated to be 13.6575, 3.6544, 0.9778 and 0.2616 $\mu lO_2\ mg^{-1}.\ h^{-1}$ respectively.

b. Bimodal oxygen uptake from closed respirometer at 22.5 ± 1.0⁰C: Here the oxygen uptake was simultaneously measured from still water as well as from air:

Oxygen uptake per unit time from the air: Oxygen uptake of the early developmental stages of *Anabas testudineus* through the accessory respiratory organs alone ranged from 4.10±0.16 to 73.30±0.30 $\mu lO_2.fish^{-1}.\ h^{-1}$ within the body weight range of 56.00±2.00 to 1026.50±6.50 mg (Table – 2). On log/log co-ordinates, the data showed a straight-line relationship with a slope of 1.0019 and intercept (=a) 0.0612 (Table – 8). It suggests that aerial oxygen uptake increases with increasing body weight by a power of 1.0019. The two variables showed a highly positive correlation ($r=0.9930$, $p<0.001$). The following relationship is applicable:

$$\dot{V}O_2 = 0.0612\ W^{1.0019}$$

$$\text{Or, Log } \dot{V}O_2 = -1.2136 + 1.0019 \times \log W$$

$$0.9229 < b < 1.0808$$

Oxygen uptake per unit body weight from the air: The weight-specific aerial oxygen uptake increases with increasing body weight (56.00 ± 2.00 to 1026.50 ± 6.50 mg) and ranged from 0.0732 ± 0.009 to 0.0715 ± 0.0001 $\mu lO_2\ mg^{-1}.h^{-1}$ (Table – 3). When these data were plotted on log/log coordinates, they gave a straight line with the slope of 0.0026 and intercept 0.0609. The correlation between the two variables was significant ($r = 0.0215$, $p<0.1$) (Table – 9). The relationship may be expressed as below:

$$\dot{V}O_2 = 0.0609W^{0.0026}$$

$$\text{Or, Log } V\dot{O}_2 = -1.2152 + 0.0026 \times \log W$$

$$-0.0765 < b < 0.0816$$

Oxygen uptake per unit time from still water: Oxygen uptake through gills alone increases with an increase in similar bodyweight and it ranged from 21.20 ± 0.20 to 61.30 ± 0.16$\mu 1O_2.\ fish^{-1}.h^{-1}$ (Table – 2). When data were plotted on log/ log co-ordinates, they depicted a straight line with a slope of 0.3655 and intercept 4.3904 (Table – 8). Both the variables are highly correlated ($r = 0.9493$, $p < 0.001$) and can be shown by the following equations:

$\dot{V}O_2 = 4.3904W^{0.3655}$

Or, $\text{Log } \dot{V}O_2 = 0.6425 + 0.3655 \times \log W$

$0.2851 < b < 0.4458$

Oxygen uptake per unit body weight from still water: Weight specific oxygen uptake through gills showed a decreasing trend from 0.3788 ± 0.0099 to 0.0598 ± 0.0002 $\mu lO_2\ mg^{-1}.\ h^{-1}$ with an increase in body weight from 56.00 ± 2.00 to 1026.50 ± 6.50 mg (Table – 3). When these data were plotted on log/ log co-ordinates, they gave a straight line with a slope of -0.6351 and the intercept being 4.4059 (Table – 9). The two variables showed a close and highly significant correlation ($r = 0.9821, p < 0.001$). Therefore, the relationship between these two variables can be represented by the following equations:

$\dot{V}O_2 = 4.4059\ W^{-0.6351}$

Or, $\text{Log } \dot{V}O_2 = 0.6440 - 0.6351 \times \log W$

$-0.7160 < b < -0.5543$

Total oxygen uptake per unit time: The total oxygen uptake (aerial + aquatic) also showed an increasing trend from 25.30 ± 0.36 to 134.60 ± 0.46$\mu lO_2.fish^{-1}.\ h^{-1}$ with an increase in bodyweight from 56.00 ± 2.00 to 1026.50 ± 6.50 mg (Table–2). Plotting of the bilog graph of these two variables gave a straight line with a slope of 0.5801 and intercept to be 2.0460 (Table – 8). The two variables showed a close and positive correlation ($r = 0.9758, p < 0.001$). Thus, the relationship between the above two variables can be shown in logarithmic form as:

$\dot{V}O_2 = 2.0460\ W^{0.5801}$

Or, $\text{Log } \dot{V}O_2 = 0.3109 + 0.5801 \times \log W$

$0.4939 < b < 0.6664$

Total oxygen uptake per unit body weight: The weight-specific total oxygen uptake ranged from 0.4520 ± 0.0108 to 0.1313 ± 0.0003 $\mu lO_2\ mg^{-1}.\ h^{-1}$ within body weight range of 56.00 ± 2.00 to 1026.50 ± 6.50 mg (Table – 3). On log/ log co-ordinates, the data gave a straight-line relationship with a slope and intercept of -0.4200 and 2.0478 respectively (Table – 9). The high correlation ($r = 0.9550, p < 0.001$) was shown by these two variables. The relationship between the body weight and oxygen uptake may be represented in the following logarithmic form:

$\dot{V}O_2 = 2.0478\ W^{-0.4200}$

Or, $\text{Log}\ VO_2 = 0.3113 - 0.4200 \times \log\dot{W}$

$0.5065 < b < 0.3334$

The average aerial, aquatic and total oxygen uptake per unit time was found to be 33.94 ± 22.25,41.83 ± 13.33 and 75.77 ± 34.99μlO_2. fish^{-1}. h^{-1} respectively and its corresponding weight-specific oxygen uptake came as 0.0622 ± 0.0064,0.1141 ± 0.0901 and 0.1762 ± 0.884μlO_2 mg^{-1}.h^{-1} respectively for a fish of average body weight of 538.35 ± 321.86 mg (Table-2 and 3). The fish consumes about 40.21 ± 11.97 and 59.79 ± 11.97% oxygen respectively from air and still water. The average aquatic/ aerial respiration for this fish was calculated to be1.827 ± 1.312. The value of aerial, aquatic and total oxygen uptake per unit time (μlO_2.fish^{-1}. h^{-1}) and per unit body weight (μlO_2 mg^{-1}. h^{-1}) for a fish of 1.0, 10.0, 100.0 and 1000.0 mg body weight was calculated and it was depicted in Table–10.

B. Oxygen uptake in *Channa punctatus*

a. Aquatic oxygen uptake from continuously flowing water at 28.0±1.0°C:

Oxygen uptake per unit time (μlO_2.fish^{-1}.h^{-1}): With the gradual increase in body weight from 8.22 ±3.89 to 925.50 ±3.54mg, the rate of aquatic oxygen uptake increased from 22.482 ± 4.936 to 226.927 ± 1.556 μlO_2 fish^{-1}.h^{-1}(Table – 4). When linear regression of logarithmic of oxygen uptake of corresponding fish was calculated, the following equation was obtained with r=0.9632. The relation was found to be highly significant ($p < 0.001$) (Table – 7).

$\dot{V}O_2 = 13.8801\ W^{0.4064}$
Or, $\text{Log}\ \dot{V}O_2 = 1.1424 + 0.4064 \times \log W$
$0.3357 < b < 0.4771$

Table 2: Rate of bimodal oxygen uptake (µlO2.h-1) in relation to body weight during early developmental stages of *Anabas testudineus* at 22.5±1.0°C.

S. No.	Body weight of a fish (mg)	Bimodal Oxygen Uptake (µlO$_2$.h^{-1})						Aerial (%)	Aquatic (%)
		Aerial		Aquatic		Total			
		Calculated (Estimated)	95% CL	Calculated (Estimated)	95% CL	Calculated (Estimated)	95% CL		
1.	56.00 ±2.00	04.10±0.16 (3.4504)	2.5791 4.6162	21.20±0.20 (19.1148)	14.2134 25.7065	25.30±0.36 (21.7892)	15.3831 24.0500	16.20	83.79
2.	125.00±5.00	06.50±0.25 (7.7133)	5.9356 10.0234	26.50±0.21 (26.6341)	19.6340 33.4680	33.00±0.46 (33.6821)	25.3016 44.8380	19.69	80.30
3.	225.00±2.50	14.70±0.30 (13.8991)	10.8366 17.8281	31.40±0.20 (30.2752)	24.6652 40.9393	46.10±0.50 (47.3687)	36.0948 62.1640	31.89	68.11
4.	330.00±3.00	21.40±0.40 (20.3998)	15.9707 26.0572	33.00±0.30 (36.5513)	28.4903 46.8928	54.40±0.70 (59.1542)	45.2789 77.2810	39.34	60.66
5.	350.00±4.00	19.80±0.20 (21.6385)	16.9457 27.6309	32.60±0.24 (37.3458)	29.1188 47.8971	52.40±0.44 (61.2083)	46.8670 79.9380	37.79	62.21
6.	465.00±0.00	25.60±0.16 (28.7634)	22.5310 36.7199	37.40±0.30 (41.4319)	32.3129 53.1243	63.00±0.46 (72.1756)	55.2800 94.2350	40.63	59.37
7.	505.00±5.00	29.20±0.20 (31.2424)	24.4652 39.8972	40.20±0.24 (42.7005)	33.2916 54.7683	69.40±0.44 (75.7150)	57.9704 98.8900	42.07	27.92
8.	600.00±3.30	34.80±0.30 (37.2424)	29.0412 47.4752	45.20±0.30 (45.4770)	35.4127 58.4016	80.00±0.60 (83.6778)	63.9824 109.4350	43.50	56.50
9.	616.00±5.00	36.80±0.16 (38.1236)	29.8101 48.7559	43.80±0.50 (45.9166)	35.7458 58.9810	80.60±0.66 (84.9652)	64.9490 111.1490	45.66	54.34
10.	800.00±4.08	49.00±0.16 (49.5351)	38.6100 63.5669	64.20±0.20 (50.6101)	39.1920 65.1190	113.20±0.36 (98.8762)	75.3012 129.8310	43.29	56.71
11.	933.10±3.01	62.00±0.24 (57.8054)	44.9204 74.3869	55.80±0.30 (53.4400)	41.3454 69.0877	117.80±0.54 (108.1237)	82.0936 142.4070	52.63	47.37
12.	966.40±1.27	64.00±0.41 (59.8718)	46.4931 77.1008	51.20±0.33 (54.1352)	41.8483 70.0293	115.20±0.75 (110.3453)	83.7153 145.4460	55.56	44.44

S. No.	Body weight of a fish (mg)	Bimodal Oxygen Uptake ($\mu lO_2.h^{-1}$)						Aerial (%)	Aquatic (%)
		Aerial		Aquatic		Total			
		Calculated (Estimated)	95% CL	Calculated (Estimated)	95% CL	Calculated (Estimated)	95% CL		
13	1026.50±6.50	73.30±0.30 (63.5954)	49.3210 82.0014	61.30±0.16 (55.3399)	42.7235 71.6816	134.60±0.46 (114.2686)	86.5700 150.8290	54.46	45.54
Average	538.35±321.86	33.94±22.25 (33.3207±19.9483)		41.83±13.33 (41.4586± 11.1936)		75.77±34.99 (74.7192± 29.4165)		40.21± 11.97	59.79± 11.97

Table 3: Rate of bimodal oxygen uptake ($\mu lO_2.mg^{-1}.h^{-1}$) in relation to body weight during early developmental stages of *Anabas testudineus* at 22.5±1.0°C.

S.No.	Body weight of a fish (mg)	Bimodal Oxygen Uptake ($\mu lO_2.mg^{-1}.h^{-1}$)						Aerial (%)	Aquatic (%)
		Aerial		Aquatic		Total			
		Calculated (Estimated)	95% CL	Calculated (Estimated)	95% CL	Calculated (Estimated)	95% CL		
1.	56.00 ±2.00	0.0732±0.0009 (0.0616)	0.0277 0.1370	0.3788±0.0099 (0.3445)	0.1512 0.7850	0.4520±0.0108 (0.3777)	0.1573 0.9066	16.19	83.80
2.	125.00±5.00	0.0520±0.0001 (0.0617)	0.0260 0.1463	0.2172±0.0065 (0.2064)	0.0848 0.5020	0.2642±0.0066 (0.2696)	0.1048 0.6937	19.68	80.21
3.	225.00±2.50	0.0653±0.0006 (0.0618)	0.0249 0.1535	0.1400±0.0002 (0.14818)	0.0556 0.3618	0.2053±0.0008 (0.2106)	0.0778 0.5702	31.81	68.19
4.	330.00±3.00	0.0648±0.0007 (0.0618)	0.0241 0.1584	0.1000±0.0000 (0.1110)	0.0422 0.2923	0.1648±0.0007 (0.1793)	0.0641 0.5019	39.32	60.68
5.	350.00±4.00	0.0566±0.0002 (0.0618)	0.0240 0.1591	0.0931±0.0002 (0.1070)	0.0404 0.2829	0.1497±0.0004 (0.1749)	0.0622 0.4921	37.81	62.19
6.	465.00±0.00	0.0551±0.0001 (0.0619)	0.0235 0.1628	0.0803±0.0003 (0.0892)	0.0330 0.2415	0.1354±0.0004 (0.1553)	0.0538 0.4476	40.69	59.31
7.	505.00±5.00	0.0578±0.0002 (0.0619)	0.0234 0.1640	0.0796±0.0002 (0.0846)	0.0311 0.2307	0.1374±0.0004 (0.1500)	0.0516 0.4355	42.07	57.93
8.	600.00±3.30	0.0580±0.0001 (0.0619)	0.0231 0.1663	0.0753±0.0000 (0.0758)	0.0274 0.2095	0.1333±0.0001 (0.1395)	0.0473 0.4112	43.51	56.49
9.	616.00±5.00	0.0598±0.0002 (0.0619)	0.0230 0.1666	0.0708±0.0003 (0.0746)	0.0269 0.2065	0.1306±0.0005 (0.1380)	0.0467 0.4067	45.79	54.21

S.No.	Body weight of a fish (mg)	Bimodal Oxygen Uptake ($\mu lO_2.mg^{-1}.h^{-1}$)						Aerial (%)	Aquatic (%)
		Aerial		Aquatic		Total			
		Calculated (Estimated)	95% CL	Calculated (Estimated)	95% CL	Calculated (Estimated)	95% CL		
10.	800.00±4.08	0.0613±0.0001 (0.0620)	0.0226 0.1702	0.0803±0.0002 (0.0631)	0.0233 0.1785	0.1416±0.0003 (0.1236)	0.0409 0.3736	43.29	56.71
11.	933.10±3.01	0.0662±0.0001 (0.0620)	0.0223 0.1724	0.0597±0.0003 (0.0572)	0.2000 0.1639	0.1260±0.0004 (0.1159)	0.0378 0.3548	52.62	47.38
12.	966.40±1.27	0.0666±0.0010 (0.0620)	0.0222 0.1729	0.0530±0.0004 (0.0559)	0.0195 0.1607	0.1196±0.0014 (0.1142)	0.0372 0.3508	55.59	44.31
13	1026.50±6.50	0.0715±0.0001 (0.0620)	0.0221 0.1737	0.0598±0.0002 (0.0538)	0.0186 0.1554	0.1313±0.0003 (0.1113)	0.0361 0.3438	54.46	45.54
Average	538.35±321.86	0.0622±0.0064		0.1141±0.0901		0.1762±0.0884		40.22± 11.52	59.77± 11.51

$\dot{V}$The oxygen uptake of a fish 1.0, 10.0, 100.0 and 1000.0 mg body weight was calculated to be 13.880, 35.381, 90.190 and 229.901 μlO_2 fish^{-1}. h^{-1} respectively (Table – 10). Therefore, with a unit increase in body weight, the rate of aquatic oxygen uptake through gills increases by a power function of 0.4064 A *Channa punctatus* of an average body weight of 368.694 ± 311.317 consumes 135.988±54.466 μlO_2 fish^{-1}. h^{-1}.

Oxygen uptake per unit body weight (μlO_2 mg^{-1}. h^{-1}): *Channa punctatus* weighing 8.22 ± 3.89 to 925.50 ± 3.54 mg shows decrease in weight-specific oxygen uptake from 3.107 ± 1.166 to 0.246 ± 0.001μlO_2 mg^{-1}. h^{-1} (Table – 4). When data on oxygen uptake through gills were plotted against body weight on the biolograthmic grid, they depicted a slope of -0.6141 and intercept 15.6365 (Table-7). Therefore, estimated value of $\dot{V}O_2$ through gills for 1.0, 10.0, 100.0 and 1000.0 mg fish was calculated to be 15.637, 3.803, 0.925 and 0.225 μlO_2 mg^{-1}. h^{-1} respectively (Table-10). The relationship between the two variables is:

$\dot{V}O_2 = 15.6365\ W^{-0.6141}$

Or, Log $\dot{V}O_2 = 1.1941\text{-}0.6141 \times \log W$

$-0.6780 < b < -0.5501$

The correlation was negative and highly significant ($r=0.9866$, $p<0.001$). The observed slope (-0.6141) was slightly less than the estimated slope (-0.5936).

b. Bimodal oxygen uptake from closed respirometer at 22.5 ± 1.0^0 C: The oxygen uptake per unit time (μlO_2. fish^{-1}. h^{-1}) simultaneously from the air and still water gradually increases but weight-specific oxygen uptake (μlO_2.mg^{-1}.h^{-1}) decreases when a fish was kept in normoxic water with free access to air:

Oxygen uptake per unit time from the air: Oxygen uptake from the air through air-breathing organs (ABOs) showed an increasing trend from 3.80 ± 0.24 to 45.00 ± 0.30 μlO_2.fish^{-1}. h^{-1} with an increase in body weight from 60.00 ± 3.01 to 1040.00 ± 3.50 mg (Table – 5). The two variables showed a close and highly significant correlation ($r=0.9793$, $p<0.001$). When these data were plotted on log/ log co-ordinates, they gave a straight line with the slope of 0.9533 and intercept being 0.0647 (Table-8). Therefore, the relationship between these two variables can be represented by the following equations.

$\dot{V}O_2 = 0.0467\ W^{0.9533}$
Or, Log $\dot{V}O_2 = -1.1890 + 0.9533 \times \log W$
$-0.8048 < b < 1.1017$

Table 4: Rate of aquatic oxygen uptake ($\mu lO_2.fish^{-1}.h^{-1}$ and $\mu l\ O_2.mg^{-1}.h^{-1}$) in relation to body weight during early developmental stages of *Channa punctatus* at 28.0±1.0°C.

S. No.	Body weight of a fish (mg)	Aquatic Oxygen Uptake						$\mu lO_2.g^{-1}.h^{-1}$ Estimated
		$\mu lO_2.fish^{-1}.h^{-1}$			$\mu lO_2.mg^{-1}.h^{-1}$			
		Calculated	Estimated	95% CL for estimated value	Calculated	Estimated	95% CL for estimated value	
1.	8.22±3.89	22.482±4.936	32.6756	21.1345 50.5185	3.107±1.166	4.2883	2.9017 6.3374	2734.37
2.	31.67±8.74	69.227±8.235	56.5215	38.1246 83.7951	2.210±0.094	1.8736	1.3164 2.6666	2186.09
3.	57.17±8.42	94.592±2.163	71.8561	49.0604 105.2428	1.655±0.025	1.3036	0.9259 1.8353	1654.69
4.	89.25±9.55	101.244±4.385	86.1165	59.1677 125.3388	1.135±0.068	0.9916	0.7083 1.3882	1134.39
5.	131.08±8.92	110.699±3.896	100.6767	69.4017 146.0443	0.888±0.069	0.7831	0.5611 1.0931	844.49
6.	187.92±2.38	117.752±2.916	116.5448	80.4446 168.8438	0.627±0.008	0.6277	0.4503 0.8752	626.62
7.	229.13±5.07	121.937±2.247	126.3240	87.1910 183.0196	0.545±0.034	0.5558	0.3986 0.7749	532.18
8.	353.13±4.42	134.060±0.086	150.6008	103.7433 218.6206	0.376±0.009	0.4261	0.3051 0.5952	379.64
9.	459.00±2.00	163.850±3.830	167.5350	115.1302 243.7941	0.357±0.009	0.3628	0.2592 0.5078	356.97

S. No.	Body weight of a fish (mg)	Aquatic Oxygen Uptake						$\mu lO_2.g^{-1}.h^{-1}$ Estimated
		$\mu lO_2.fish^{-1}.h^{-1}$			$\mu lO_2.mg^{-1}.h^{-1}$			
		Calculated	**Estimated**	**95% CL for estimated value**	**Calculated**	**Estimated**	**95% CL for estimated value**	
10.	529.67±5.51	166.619±3.592	177.5747	121.8206 258.8438	0.300±0.023	0.3322	0.2370 0.4657	314.57
11.	634.00±5.66	175.263±2.471	191.0355	130.7240 279.1604	0.276±0.006	0.2975	0.2117 0.4180	276.44
12.	704.00±9.00	185.495±0.867	199.3415	136.1823 291.7905	0.262±0.004	0.2790	0.1983 0.3925	263.49
13	822.00±6.00	213.680±1.169	212.2985	144.6414 311.6000	0.258±0.002	0.2538	0.1798 0.3578	259.95
14.	925.50±3.54	226.927±1.556	222.7806	151.4381 327.7299	0.246±0.001	0.2358	0.1668 0.3333	245.19
Average	368.694± 311.317	135.988± 54.466	136.563± 61.168		0.876±0.875	0.9008 ± 1.083		843.506 ± 801.569

Oxygen uptake per unit body weight from air: The weight-specific aerial oxygen uptake decreases with a similar increase in body weight and it ranged from 0.0633±0.0001 to 0.0433±0.0001 $\mu lO_2 . mg^{-1}.h^{-1}$ (Table -6). When data were plotted on log/ log co-ordinates, they gave a straight line with a slope of -0.0466 and intercept 0.0647 (Table – 9). Both the variables are negative correlated ($r=0.2305$, $p<0.5$). The relationship between the two variables can be shown by the following equations:

$\dot{V}O_2 = 0.0647\ W^{-0.0466}$

Or, $\text{Log}\ \dot{V}O_2 = -1.1893 - 0.0466 \times \log\dot{W}$

$-0.1950<b<-0.1018$

Oxygen uptake per unit time from still water: The aquatic oxygen uptake through the gills alone ranged from 24.20±0.30 to 72.60 ± 0.50 $\mu lO_2.fish^{-1}.h^{-1}$ within the bodyweight range of 60.00±3.01 to 1040.00±3.50mg (Table -5). On log/ log co-ordinates, the data showed a straight-line relationship with a slope and intercept of 0.3625 and 5.0050 respectively (Table -8). A high correlation ($r=0.9441$, $p<0.001$) was shown by these two variables. The following logarithmic form may represent the relationship between the body weight and oxygen uptake:

$\dot{V}O_2 = 5.0050\ W^{0.3625}$

Or, $\text{Log}\ \dot{V}O_2 = 0.6996 + 0.3625 \times \log W$

$0.2671 < b < 0.4579$

Oxygen uptake per unit body weight from still water: The weight-specific oxygen uptake from still water showed a decreasing trend from 0.4033±0.0003 to 0.0698±0.0003$\mu lO_2.mg^{-1}.h^{-1}$ with an increase in body weight from 60.00±3.01 to 1040.00±3.50 mg (Table – 6). The log/ log graph of these two variables gave a straight line with a slope of –0.6368 and intercept being 4.9777 (Table-9). The two variables showed a close and negative correlation ($r=0.9808$, $p<0.001$). Thus, the relationship between the above two variables can be shown in logarithmic form as:

$\dot{V}O_2 = 4.9777W^{-0.6368}$

or, $\text{Log}\ \dot{V}O_2 = 0.6970 - 0.6368 \times \log W$

$-0.7321 < b < -0.5414$

Total oxygen uptake per unit time: Total oxygen uptake through both air – breathing organs and gills also increased with increasing body weight and ranged from 28.00±0.54 to 117.60±0.80 $\mu lO_2.fish^{-1}.h^{-1}$ (Table -5). When these data were plotted on log/ log co-ordinates, they gave a straight line with a slope of 0.5187 and an intercept of 2.9255 (Table-8). The correlation ($r=0.9861$) between the two variables is highly significant ($p<0.001$). In logarithmic form, these two variables are represented as:

$$\dot{V}O_2 = 2.9255 W^{0.5187}$$

$$\text{or, } \log \dot{V}O_2 = 0.4662 + 0.5187 \times \log W$$

$$0.4528 < b < 0.5847$$

Total oxygen uptake per unit body weight: The weight-specific total oxygen uptake of the early developmental stages of *Channa punctatus* ranged from 0.4667 ± 0.0004 to $0.1131 \pm 0.0004 \mu l O_2 . mg^{-1} . h^{-1}$ within the bodyweight range of 60.00 ± 3.01 to 1040.00 ± 3.50 mg (Table-6). On log/ log co-ordinates, the data gave a straight-line relationship with a slope of -0.4840 and intercept 2.9101 (Table – 9). The two variables showed a highly significant negative correlation ($r=0.9837$, $p<0.001$). The following relationship is applicable:

$$\dot{V}O_2 = 2.9101 W^{-0.4840}$$

$$\text{Or, } \log \dot{V}O^2 = 0.4639 - 0.4840 \times \log W$$

$$-0.5468 < b < -0.4141$$

Table 5: Rate of bimodal oxygen uptake ($\mu lO_2.h^{-1}$) in relation of body weight during early developmental stages of *Channa punctatus* at 22.5±1.00 C.

S. No.	Body weight of a fish (mg)	Bimodal Oxygen Uptake ($\mu lO_2.h^{-1}$)						Aerial (%)	Aquatic (%)
		Aerial		Aquatic		Total			
		Calculated (Estimated)	95% CL	Calculated (Estimated)	95% CL	Calculated (Estimated)	95% CL		
1.	60.03±3.01	03.80±0.24 (3.206)	1.9127 5.3748	24.20±0.30 (22.077)	15.8390 30.7722	28.00±0.54 (24.469)	19.4500 30.7800	13.57	86.43
2.	140.00±2.00	05.50±0.25 (7.191)	4.5436 11.3808	30.20±0.21 (30.014)	22.3435 40.3168	35.70±0.46 (37.975)	30.9667 46.5700	15.41	84.59
3.	240.00±4.00	11.30±0.16 (12.020)	7.7597 18.6211	35.60±0.20 (36.489)	27.5409 48.3447	46.90±0.36 (50.226)	41.3481 61.0100	24.09	75.91
4.	350.00±5.00	20.60±0.30 (17.224)	11.1933 26.5023	35.60±0.24 (41.837)	31.7136 55.1905	56.20±0.54 (61.083)	50.4371 73.9800	36.65	63.34
5.	455.00±5.00	22.70±0.16 (22.118)	14.3800 34.0187	41.90±0.50 (46.010)	34.8869 60.6801	64.60±0.66 (69.989)	57.8013 84.7500	35.14	64.86
6.	520.00±2.50	18.20±0.40 (25.120)	16.3134 38.6806	55.80±0.30 (48.292)	36.5902 63.7358	74.00±0.70 (75.009)	61.9156 90.8700	24.59	75.40
7.	620.00±3.20	36.50±0.30 (29.706)	19.2359 45.8731	46.50±0.32 (51.471)	38.9269 68.0571	83.00±0.62 (82.174)	67.7439 99.6800	43.98	56.02
8.	707.00±3.00	36.50±0.20 (33.666)	21.7323 52.1541	49.30±0.24 (53.980)	40.7418 71.5192	85.80±0.44 (87.967)	72.4176 106.8500	42.54	57.45
9.	815.00±5.00	40.10±0.16 (38.552)	24.7780 59.9834	54.40±0.24 (56.834)	42.7762 75.5120	94.50±0.40 (94.699)	77.8089 115.2500	42.43	57.57

S. No.	Body weight of a fish (mg)	Bimodal Oxygen Uptake ($\mu lO_2.h^{-1}$)						Aerial (%)	Aquatic (%)
		Aerial		Aquatic		Total			
		Calculated (Estimated)	95% CL	Calculated (Estimated)	95% CL	Calculated (Estimated)	95% CL		
10.	930.00±4.00	45.00±0.20 (43.722)	27.9630 68.3603	67.00±0.33 (59.620)	44.7314 79.4624	112.00±0.53 (101.410)	83.1420 123.6900	40.18	59.82
11.	1040.00±3.50	45.00±0.30 (48.638)	30.9594 76.4112	72.60±0.50 (62.085)	46.4385 83.0025	117.60±0.80 (107.464)	87.9193 131.3500	38.26	61.74
Average	534.27 ± 321.84	25.93 ± 15.40 (25.560 ± 14.887)		46.65±15.09 (46.246 ± 12.631)		72.57±29.47 (72,042 ± 26.495)		32.71± 10.61	67.56 ± 11.09

The values for aerial, aquatic and total oxygen uptake in terms of per unit time (μlO_2. fish^{-1}. h^{-1}) and per unit body weight (μlO_2. mg^{-1}. h^{-1} for a fish of 1.0, 10.0, 100.0 and 1000.0 mg were also calculated and arranged in Table –

10. The average aerial, aquatic and total oxygen uptake per unit time was calculated to be 25.93 ± 15.40, 45.65 ± 15.09 and 72.57 ± 29.47 $\mu lO_2.fish^{-1}.\ h^{-1}$ respectively and its corresponding oxygen uptake in terms of per unit body weight came as 0.0495 ± 0.0085,0.1291 ± 0.1013 and 0.1787 ± 0.1043 $\mu lO_2.\ mg^{-1}.\ h^{-1}$ respectively for a fish of average body weight 534.27 ± 321.84 mg (Tables 5 and 6). The fish consumes about 32.71 ± 10.61% and 67.56 ± 11.09% oxygen from the air and still water respectively. The average ratio of aquatic/ aerial respiration for a fish of average body weight of 534.27±321.84 mg was calculated to be 2.612 ± 1.77.

Table 6: Rate of bimodal oxygen uptake ($\mu lO_2 . mg^{-1}.h^{-1}$) in relation to body weight during early developmental stages of *Channa punctatus* at 22.5± 1.0°C.

S.No.	Body weight of a fish (mg)	Bimodal Oxygen Uptake ($\mu lO_2 . mg^{-1} h^{-1}$)						Aerial (%)	Aquatic (%)
		Aerial		Aquatic		Total			
		Calculated (Estimated)	95% CL	Calculated (Estimated)	95% CL	Calculated (Estimated)	95% CL		
1.	60.03±3.01	0.0633±0.0001 (0.0534)	0.0319 0.0896	0.4033±0.003 (0.3671)	0.2634 0.5116	0.4667±0.0004 (0.4070)	0.3231 0.5127	13.56	86.42
2.	140.00±2.00	0.0393±0.0009 (0.0514)	0.0325 0.0813	0.2143±0.0099 (0.2140)	0.1594 0.2875	0.2536±0.0106 (0.2709)	0.2207 0.3326	15.50	84.50
3.	240.00±4.00	0.0471±0.0001 (0.5001)	0.0323 0.0776	0.1483±0.0002 (0.1519)	0.1146 0.2012	0.1954±0.0003 (0.2091)	0.1720 0.2543	24.10	75.89
4.	350.00±5.00	0.0589±0.0001 (0.0492)	0.0320 0.0757	0.1017±0.0065 (0.1194)	0.0905 0.1575	0.1606±0.0066 (0.1744)	0.1439 0.2115	36.67	63.33
5.	455.00±5.00	0.0499±0.0002 (0.0486)	0.0316 0.0748	0.0921±0.0003 (0.1011)	0.0766 0.1333	0.1420±0.0005 (0.1538)	0.1269 0.1864	35.14	64.86
6.	520.00±2.50	0.0350±0.0006 (0.0483)	0.0314 0.0744	0.1073±0.0002 (0.0928)	0.0703 0.1225	0.1423±0.0008 (0.1442)	0.1189 0.1749	24.59	75.40
7.	620.00±3.20	0.0589±0.0001 (0.0479)	0.0310 0.0740	0.0750±0.0000 (0.0829)	0.0628 0.1097	0.1339±0.0001 (0.1325)	0.1091 0.1609	43.99	56.01
8.	707.00±3.00	0.0516±0.0007 (0.0476)	0.0308 0.0738	0.0697±0.0000 (0.0763)	0.0576 0.1011	0.1214±0.0007 (0.1244)	0.1023 0.1513	42.50	57.41

S.No.	Body weight of a fish (mg)	Bimodal Oxygen Uptake (µlO_2. mg^{-1} h^{-1})						Aerial (%)	Aquatic (%)
		Aerial		Aquatic		Total			
		Calculated (Estimated)	95% CL	Calculated (Estimated)	95% CL	Calculated (Estimated)	95% CL		
9.	815.00±5.00	0.0492±0.0002 (0.0473)	0.0304 0.0736	0.0667±0.0002 (0.0697)	0.0525 0.0926	0.1160±0.0004 (0.1162)	0.0954 0.1416	42.41	57.50
10.	930.00±4.00	0.0484±0.0002 (0.0470)	0.0301 0.0735	0.0720±0.0002 (0.0641)	0.0481 0.0854	0.1204±0.0004 (0.1091)	0.0893 0.1332	40.20	59.80
11.	1040.00±3.50	0.0433±0.0001 (0.0468)	0.0298 0.0735	0.0698±0.0003 (0.0597)	0.0446 0.0798	0.1131±0.0004 (0.1034)	0.0845 0.1265	38.28	61.72
Average	534.27 ± 321.84	0.0495 ± 0.0085		0.1291±0.1013		0.1787±0.1043		32.45± 10.57	67.53± 10.57

Table 7: Computed values of Intercept, Slope and Correlation coefficient to show the relationship between body weight and aquatic oxygen uptake ($\mu lO_2 . h^{-1}$ and $\mu lO_2 . mg^{-1} . h^{-1}$) of *Anabas testudineus* and *Channa punctatus* at 28.0 ± 1.0°C with 95% confidence limit and standard error of intercept and slope.

Fish	Body weight vs.	Intercept (=log a)			Slope			Correlation coefficient
		Estimated value	95% CL	SE	Estimated value	95% CL	SE	
Anabas testudineus	$\dot{V}O_2$ ($\mu lO_2 . h^{-1}$)	1.0758	0.9145 1.2371	0.0733	0.4522	0.3827 0.5217	.0316	0.9742 ($p<0.001$)
	$\dot{V}O_2$ ($\mu lO_2 . mg^{-1}.h^{-1}$)	1.1354	0.9859 1.2849	0.0679	-0.5726	-0.6370 -0.5082	0.0293	0.9859 ($p<0.001$)
Channa punctatus	$\dot{V}O_2$ ($\mu lO_2 .h^{-1}$)	1.1424	0.9725 1.3123	0.0780	0.4064	0.3357 0.4771	0.0327	0.9632 ($p<0.001$)
	$\dot{V}O_2$ ($\mu lO_2 .mg^{-1}.h^{-1}$)	1.1941	1.0410 1.3464	0.0699	-0.6141	-0.6780 -0.5501	0.0293	0.9866 ($p<0.001$)

Table 8: Intercept, slope and correlation coefficient showing the relationship of bimodal oxygen uptake $\mu lO_2.h^{-1}$ to their respective body weights of *Anabas testudineus* and *Channa punctatus* at confidence limit and standard error of intercept and slope are also tabulated.

Fish	Body weight vs. Oxygen uptake	Intercept (=log a)			Slope			Correlation Coefficient
		Estimated	95% CL	SE	Estimated	95% CL	SE	
Anabas testudineus	Aerial	-1.2136	-1.4224 -1.0047	0.0949	1.0019	0.9229 1.0808	0.0359	0.9930 ($p<0.001$)
	Aquatic	0.6425	0.4299 0.8550	0.0966	0.3655	0.2851 0.4458	0.0365	0.9493 ($p<0.001$)
	Total (Aerial + Aquatic)	0.3109	0.0828 0.5390	0.1036	0.5801	0.4939 0.6664	0.0391	0.9758 ($p<0.001$)
Channa punctatus	Aerial	-1.1890	-1.5809 -0.7972	0.1732	0.9533	0.8048 1.1017	0.0656	0.9793 ($p<0.001$)
	Aquatic	0.6994	0.4476 0.9513	0.1113	0.3625	0.2671 0.4579	0.0422	0.9441 ($p<0.001$)
	Total (Aerial + Aquatic)	0.4662	0.2921 0.6403	0.0770	0.5187	0.4528 0.5847	0.0292	0.9861 ($p<0.001$)

Table 9: Intercept, slope and correlation coefficient showing the relationship of bimodal oxygen uptake μlO_2. $mg^{-1}.h^{-1}$ to their respective body weights of *Anabas testudineus* and *Channa punctatus* at 22.5 ± 1.0^0 C.95% confidence limit and standard error of intercept and slope are also tabulated.

Fish	Body weight vs. Oxygen uptake	Intercept (=log a)			Slope			Correlation Coefficient
		Estimated	95% CL	SE	Estimated	95% CL	SE	
Anabas testudineus	Aerial	-1.2152	-1.4224 -1.0060	0.0950	0.0026	-0.0765 0.0814	0.0359	0.0215 (p>0.05)
	Aquatic	0.6440	0.4303 0.8578	0.0971	-0.6351	-0.7160 -0.05543	0.0367	0.9821 (p<0.001)
	Total (Aerial + Aquatic)	0.3113	0.0823 0.5403	0.1040	-0.4200	-0.5065 -0.3334	0.0393	0.9550 (p<0.001)
Channa punctatus	Aerial	-1.1893	-1.5810 -0.7976	0.1732	-0.0466	-0.1950 -0.1018	0.0656	0.2305 (p>0.05)
	Aquatic	0.6970	0.4453 0.9448	0.1113	-0.6368	-0.7321 -0.5414	0.0422	0.19808 (p<0.001)
	Totla (Aerial + Aquatic)	0.4639	0.2888 0.6390	0.0774	-0.4840	-0.5468 -0.4141	0.0293	0.9837 (p<0.001)

Table 10: Estimated values of aquatic and bimodal oxygen uptake for a fish of *Anabastestudineus* and *Channa punctatus* of body weight 1.0, 10.0, 100.0 and 1000.0 mg.

Type		Unit	*Anabas testudineus* (mg)				*Channa punctatus* (mg)			
			10.0	100.0	1000.0	1.0	10.0	100.0	1000.0	
A. Aquatic oxygen uptake at 28.0±1.0⁰C		µlO_2.h^{-1} µlO_2.mg^{-1}.h^{-1}	11.907 13.658	33.731 3.6540	95.553 0.9780	270.6880 0.2620	13.880 15.637	35.381 3.809	90.190 0.925	229.901 0.225
B. Bimodal Oxygen uptake at 22.5±1.0⁰C	Aquatic	µlO_2. h^{-1} µlO_2.mg^{-1}. h^{-1}	4.3900 4.4059	10.184 1.0206	23.627 0.2365	54.812 0.0548	5.0010 4.9777	11.532 1.1489	26.568 0.2652	61.2090 0.0612
	Aerial	µlO_2 O_2.h^{-1} µlO_2.mg^{-1}.h^{-1}	0.0620 0.0609	0.6140 0.0613	6.1680 0.0617	61.944 0.0620	0.0650 0.0647	0.5810 0.0581	5.2190 0.0522	46.8530 0.0469
	Total	µlO_2.h^{-1} µlO_2.mg^{-1}.h^{-1}	2.0460 2.0478	7.7810 0.7786	29.592 0.2960	112.541 0.1126	2.9260 2.9101	9.6600 0.9626	31.893 0.3184	105.3000 0.1053

Discussion

The relationship between body weight and oxygen uptake has been a subject of considerable interest and discussion for many years. Oxygen uptake of an organism in unit time is defined as metabolic rate and it represents the intensity of metabolism when related to unit body weight. Rüttner (1926) has stated that the rate of oxygen uptake of many organisms is doubled with each rise of 10°C (=283K) within the physiological limits. According to Fry and Hart (1948), the standard metabolic rate (SMR) of goldfish increases four times due to exercise and activity. In the animal kingdom, different slope (=b) values have been reported to the relationship between body weight and oxygen uptake. In different animals, slope value varies from 0.45 to 1.00 but usually a little more than 0.70 in most animals (Zeuthen, 1953; Winberg, 1956 and Hemingsen, 1960), Jager and Dekkers (1975) proposed 0.82 to be the average slope value to calculate data for the species whose slope values are unknown. These data, however, are useful only for water breathing fishes. The air-breathing fishes show wide variation in slope values under different conditions. Attempts have been made to evaluate the effect of growth on oxygen uptake in different fish species. Different slope values have been calculated for aquatic (surfacing prevented i.e., without any access to air and surfacing allowed or with access to air) and bimodal (simultaneous oxygen uptake through gills and air-breathing organs) oxygen uptake (Tables: 11 and 12).

The oxygen uptake during the early developmental stages of *Anabas testudineus* and *Channa punctatus* in aquatic and bimodal breathing conditions give a straight line when their body weight and oxygen uptake were plotted on log/log graphs. The relationship of metabolic rate and its intensity to the body weight in these fishes was similar to that as found in other teleosts.

Aquatic Oxygen Uptake: The surfacing prevented aquatic oxygen uptake in continuously flowing water in the early developmental stages of *Anabas testudineus* and *Channa punctatus* showed similar relationships. In *Anabas testudineus* (body weight: 9.56 to 834.83 mg), the oxygen uptake increased by a power of 0.452 while in *Channa punctatus* (body weight: 8.22 to 925.50 mg), it increased by a power function of 0.406. These slope values were found to be lower than the slope values of early developmental stages of some purely aquatic breathing teleosts (Table: 11). The lower slope value of *Anabas testudineus* and *Channa punctatus* may be attributed to the stress due to surfacing prevented conditions during experimentation. Jha (1995) observed that the oxygen uptake during early developmental stages of *Labeo rohita* (body weight: 5.00 to 930.00 mg) and *Catla catla* (body weight: 10.00 to 1000.00 mg) increased by a functional power of 0.538 and 0.561 respectively, but in adult *Labeo rohita* (body weight: 16.00 to 425.00 g) and *Catla catla* (body weight: 6.50 to 240.00 g) the increase was reported to be 0.8515 and 0.7469 respectively by Kunwar et al, (1989). While working with *Labeo rohita* of body weight 5.00 to 260.00 mg, Pandey (1988) also observed that oxygen uptake in this fish increases by 0.5976 but the slope becomes 0.8576 for adult *Labeo rohita* (16.00 to 425.00g) (Table: 11). In *Heteropneustes fossilis* oxygen uptake through gills of juveniles (up to 25.00g) and adults (above 25.00g) increased by the functional power of 0.478 and 0.799 respectively (Munshi et al, 1978). The present work thus corroborates with the above-mentioned earlier works of Munshi et al, (1978); Pandey (1988); Kunwar et al, (1989) and Jha (1995). A slope of 0.526 and 0.629 respectively have been reported by Munshi and Dube (1973) for *Anabas testudineus* (body weight: 8.00 to 65.00g) and by Hakim et al, (1983) for *Channa punctatus* (body weight: 8.33 to 134.00g). These values are more than the values obtained in the present work for *Anabas testudineus* (b=0.452) and *Channa punctatus* (b=0.406) (Table: 11). The differences are probably due to the effect of different stages of the life cycle (Kamler, 1976), their activity, different growth pattern, physiology of juveniles and adults (Schurmann and Steffensen, 1997) and changes in feeding behaviour (Smith, 1935a and b) etc.

The gill surface area of a fish is considered to be directly proportional to oxygen uptake. The intercept or the value of 1g body weight of *Anabas testudineus* was calculated to be 555.9mm^2 (Hughes et al, 1973) and for *Channa punctatus*, it was 470.39mm^2 (Hakim et al, 1978) which shows that *Anabas tesiudineus* is a more active fish than that of *Channa punctatus*. These works also support the findings of the present investigation.

As the slope values were less than one, the intensity of metabolism (=weight specific oxygen uptake) tends to fall with an increase in body weight. Theoretically, the rate of decrease of oxygen uptake can be calculated by subtracting the slope value from one. The weight-specific oxygen uptake decreased by a power of -0.573 and -0.614 respectively in *Anabas testudineus* and *Channa punctatus* (Table: 7). The decrease in weight-specific oxygen uptake among fishes may be related to the economy of food utilization, structural constraints, changes in skeletal structures and opportunities for low cost based on structural changes with size (Jones, 1972; Schmidt-Nielson, 1984).

Bimodal Oxygen Uptake: Most of the air-breathing fishes live and flourish in the adverse environmental conditions of the swamps. The efficiency of bimodal oxygen uptake and the relative dependence of the fish on gills and air-breathing organs have been many studied in many Indian species (Hughes and Singh, 1970b, 1971; Singh and Hughes, 1971, 1973; Ojha et al, 1978). The regression analysis of the investigation indicates that the slope (=b) value of bimodal oxygen uptake during the early developmental stages through water and air increased respectively by a functional power of 0.365 and 1.002 in *Anabas testudineus* and by 0.362 and 0.953 in *Channa punctatus* (Table: 8). However, in adult *Anabas testudineus* and *Channa punctatus*, oxygen uptake from water and air increased by 0.851 and 1.009 (Ghosh and Munshi, 1987) and 0.863 and 1.067 (Ghosh et al, 1990) respectively. In mature *Clarias batrachus* and *Monopterus cuchia* also, bimodal oxygen uptake through water and air increases by 0.316 and 1.069 (Patra et al, 1983) and 0.483 and 1.497 (Ghosh et al, 1980) respectively. Biswas et al, (1979) calculated the slope of oxygen uptake from still water and air to be 0.493 and 0.603 respectively in *Boleophthalamus boddaerti* (bodyweight: 2.60 to 12.79g) and suggested that in larger fishes a greater proportion of oxygen uptake must take place through air-breathing organs as the fishes grow in size. The slope values in air-breathing fishes are quite high for aerial respiration and sometimes it is more than 1. According to Zeuthen (1953), a slope of more than 1 is an exception and valid for a narrow weight range. It may be due to some factors other than body weight. In this work, some factors (such as time of day, dissolved oxygen, pH, level of light, acidity and nutrition etc.) were controlled and therefore these factors did not affect the slope value. These fishes remained in still water of the respirometer and breathe and rebreathe the same water. Thus the partial pressure of carbon dioxide in the water is gradually increased. This may induce an increase in aerial breathing (Hughes and Singh, 1970a). Obviously, in larger fishes, a larger proportion of oxygen uptake must take place through air–breathing organs. It corroborates with the work of Biswas et al, (1979), Ghosh and Munshi, (1987) and Ghosh et al, (1990).

In these fishes, oxygen uptake from still water (=closed glass respirometer) gave lower slope values (0.365 in *Anabas testudineus* and 0.362 in *Channa punctatus*) in comparison to values obtained (0.452 and 0.406 respectively in *Anabas testudineus* and *Channa punctatus*) from continuously flowing water (=cylindrical glass respirometer). While working with *Channa guchua*, Ojha et al, (1978) also reported a lower slope value (0.683) for oxygen uptake from still water in comparison to flowing water (b=0.765, Table=12). Therefore, the present work corroborates with the work of Ojha et al, (1978). It seems that the lower slope value for aquatic oxygen uptake in still water may be a result of the larger fishes depleting the respirometer more rapidly than smaller fishes thus restricting aquatic oxygen uptake and resorting to air-breathing.

The present data on the bimodal oxygen uptake of *Anabas testudineus* and *Channa punctatus* reveal that these fishes depend more on water (about 60% in *Anabus testudineus* and 68% in *Channa panctata*) than air for their total metabolic activities during early developmental stages. However, they chiefly depend on air-breathing as they grow in size. The smaller individuals of *Anabas testudineus* and *Channa punctatus* are more efficient aquatic breathers than the larger ones but with the increase in body weight, oxygen uptake from water does not increase as markedly as the aerial oxygen uptake. This is clear from the fact that an *Anabas testudineus* of average body weight of 538.35mg extracts about 40% of total oxygen uptake (75.77 $\mu lO_2.h^{-1}$) from the air, whereas, a *Channa punctatus* averaging about 534.30mg in body weight takes up 32% of the total oxygen requirement (72.57 $\mu lO_2.h^{-1}$) through aerial route. However, an adult *Anabas testudineus* (body weight: 48.71g) and *Channa punctatus* (body weight: 48.88g) consume about 55% of total oxygen uptake from the air (Ghosh and Munshi, 1987 and Ghosh et al, 1990). Ghosh and Munshi (1987) estimated that *Anabas testudineus* of 1.0g body weight extracts about 33% of total oxygen requirement through its aerial route, whereas, a *Channa punctatus* of similar body weight consumes about 36% oxygen from the atmosphere (Ghosh et al, 1990). While working on inter-specific variation in oxygen uptake of *Channa* (85.5 10.0g), Ojha et al, (1981) reported that *Channa marulius, Channa striatus, Channa gachua* and *Channa punctatus* consumed about 76, 56, 58 and 78% of oxygen through their air-breathing organs respectively.

Hughes et al, (1973) measured the surface area of gills and air-breathing organs of *Anabus testudineus*, which indicates that the increase in surface area of air-breathing organs (b=0.713) was more than the gills (b=0.615) during growth. In *Channa punctatus* also, the rate of increase in the surface area of air-breathing organs (b=0.696) was more than the gills (b=0.592) as reported by Hakim et al, (1978). Therefore, it seems that in larger *Anabas* and *Channa*, the proportion of aerial oxygen uptake will dominate over aquatic oxygen uptake. They concluded that the diffusing capacity of gills per unit body weight in *Anabas* decreases with increasing body weight and therefore larger *Anabas* are not efficient to cover the increasing oxygen demand through the gills alone so that these fishes die if not allowed to breathe air, in addition, to water breathing (Hughes et al, 1973). Similarly, the diffusing capacity ($mlO_2.min^{-1}.mm\ Hg^{-1}\ .kg^{-1}$) of air-breathing organ is higher than the gills for *Channa punctatus* (Hakim et al, 1978). Munshi et al, (1976) also observed that the gills of *Clarias batrachus* are well suited for aquatic respiration during their young stages but their efficiency decreases with an increase in body weight.

Table 11: Slope values for aquatic oxygen uptake in relation to body weight under surfacing allowed and prevented condition in some air – breathing and aquatic fishes.

Fish	Body wt. (g)	Temperature (^{0}C)	Slope of aquatic oxygen uptake		References
			Surfacing allowed (Natural breathing)	Surfacing prevented (Water breathing)	
Anabas testudineus	8.0000 – 68.000 0.0096 – 0.835	27.5±1.0 28.0±1.0	0.6661 -	0.5259 0.4522	Munshi and Dube (1973) Present author
Channa gachua		27.5±1.5	0.6810	0.6410	Ojha et al. (1978)
Channa punctatus	8.330 – 102.330 8.330 – 134.000 0.008 – 0.930	20.0±1.0 31.0±1.0 28.0±1.0	0.7872 0.6013 -	0.6249 0.6292 0.4064	Hakim et al. (1983) ,, Present author
Channa striatus	36.75 – 265.40	30.0 ± 1.0	0.5368	-	Choudhary (1979)
Clarias batrachus	15.30 – 95.70	26.5±1.0	0.8690	0.8410	Munshi et al. (1976)
Heteropneustes fossilis	9.0 – 75.0 1.0 – 25.0 26.0 – 75.0	25.0±1.0 26.0±1.0 26.0±1.0	1.8981 1.0839 0.9860	1.0646 0.4780 0.7990	Munshi et al. (1976) Munshi et al. (1978) ,,
Colisa fasciatus	2.0 – 12.0	29.0±1.5	0.6710	0.5950	Ojha et al. (1977)
Boleophthalmus boddaerti	2.3 – 13.7	26.0 ±1.0	-	0.7910	Biswas et al. (1979)
Glossogobius giuris	1 – 10 (Juvenile) 11 – 23 (Adult)	25.0±0.5	-	0.6884 0.5561	Singh and Munshi (1985) ,,
Cyprinus carpio	0.0019 – 620 up to 0.674 1.2 – 46.0	- - -	- - -	0.8320 0.9700 0.8000	Oikawa and Itazawa (1985) Kamler (1976) ,,
Labeo rohita	0.005 – 0.930 0.005 – 0.260 16.000 – 425.0	29.5±1.0 - 29.0±1.0	- - -	0.5384 0.5976 0.8515	Jha (195) Pandey (1988) Kunwar et al. (1989)
Catla catla	0.010 – 1.000 6.500 – 240.000	29.5±1.0 29.0±1.0	- -	0.5612 0.7469	Jha (1995) Kunwar et al. (1989)
Cirrhinus mrigala	5.00 – 198.00 5.25 – 180.00	21.5±1.0 30.5±1.0	- -	0.7691 0.8113	Roy and Munshi (1984) ,,

Table 12: Slope values for bimodal oxygen uptake in relation to body weight of some air–breathing fishes.

Fish	Body wt. (g)	Temperature (ºC)	Slope of bimodal oxygen uptake			Reference
			Aerial	Aquatic	Total	
Anabas testudineus	9.5 – 79.0	21.0 ±1.0	1.099	0.85	0.933	Ghosh and Munshi (1987)
	11.0 – 81.0	31.5±1.0	0.929	0.809	0.899	,,
	0.056 – 1.027	22.5±1.0	1.002	0.365	0.580	Present author
Channa gachua	3.0 – 75.0	29.0±1.5	0.454	0.603	0.536	Ojha et al. (1978)
Channa punctatus	10.5 – 91.5	21.0±1.0	1.068	0.863	0.972	Ghosh et al. (1990)
	6.5 – 87.0	31.5±1.0	0.853	0.749	0.802	,,
	0.060 – 1.040	22.5±1.0	0.953	0.362	0.519	Present author
Clarias batrachus	15.0 – 175.0	20.0±1.0	0.864	0.736	0.781	Patra et al. (1983)
	7.0 – 150.0	25.0±1.0	0.841	0.517	0.763	,,
	20.0 – 220.0	30.5±1.0	1.069	0.316	0.679	,,
Colisa fasciatus	2.50 – 8.80	31.5±1.0	0.266	0.537	0.372	Ojha et al. (1988)
Heteropneustes fossilis	16.83 – 107.16	20.0±1.0	0.819	0.839	0.833	Munshi et al. (1982)
	13 – 93	25.0±1.0	0.664	0.849	0.763	,,
	18.5 – 136.5	30.5±1.0	0.785	0.520	0.833	,,
Monopterus Cuchia	157 – 390	21.0±1.0	1.100	0.972	0.998	Ghosh (1984)
		31,5±1.0	1.497	0.483	1.264	Ghosh et. al (1980)
Boleophthalmus boddaerti	2.60 – 12.75	26.0±1.0	0.604	0.493	0.549	Biswas et al (1979)
Channa striatus	3.5 – 74.4	21.0±1.0	0.778	0.706	0.741	Kumar (2009)
	3.2 – 74.5	31.0±1.0	0.575	0.447	0.553	

Therefore, it seems that with an increase in body weight the diffusing capacity of gills decreases and the larger fishes resort to aerial breathing to satisfy their total metabolic demand.

In the present work, the quotient between aquatic and aerial respiration in the early developmental stages of *Anabas testudineus* and *Channa punctatus* was found to be 1.827 and 2.612 respectively but the quotient decreased to about 0.814 (Ghosh and Munshi, 1987) and 0.834 (Ghosh et al, 1990) in the adults of these fishes. A quotient of 0.225 and 0.325 during juveniles and adults of *Channa marulius* was obtained by Ojha et al, (1979). They explained that in *Channa marulius* air-breathing dominated over aquatic breathing in both juveniles and adult stages. Ojha et al, (1988) calculated a quotient of 0.684 in *Colisa* (=*Trichogaster*) *fasciatus* and the quotient between aquatic and aerial respiration of *Notopterus chitala* was calculated to be 0.420 by Ghosh et al, (1988). The decrease in quotient value of more than one indicates the dominance of aquatic breathing over aerial breathing during early developmental stages and the quotient less than one indicates the dominance of air-breathing over aquatic breathing. Therefore, it seems that the efficiency of oxygen uptake of the bimodal gaseous exchange machinery varies mainly with the growth of fishes besides other factors. These physiological findings corroborate the earlier embryonic and stereological findings of Hughes et al, (1986), who believed that early hatchlings of *Anabas testudineus* depend mostly on aquatic medium and in due course of development of air- breathing organ supplement the total metabolic oxygen uptake and later on dominate it.

In light of the present work, it could be inferred that the main single factor in transporting fish is giving a sufficient amount of dissolved oxygen. Heavier fish and those shipped in hotter water need more oxygen. For example, if the water temperature is increased by 10°C oxygen utilization is about multiplied two times.

4

Effect of Group Size on Oxygen Uptake in Fishes

Introduction

Oxygen uptake in fishes is affected not only by the abiotic factors of the environment but is also affected due to various biotic factors. Works on fish aggregation reveal that oxygen uptake of a fish is changed by the number of fishes mutually participating in respiration which has been demonstrated by various workers. The variation in oxygen uptake when fishes of either similar or different genera or species are kept in an association is known as the group effect. The variation in oxygen uptake due to the group effect is an established fact.

The investigation and approaches of Allee (1931); Schtefanesku (1958); Malyukina et al, (1962); Grigor'eva (1966); Umezawa and Watannbe (1979), Wares II and Igram (1979), Umezawa et al, (1981) Imabayshi and Takahashi (1987), etc. are worth mentioning and regarded as milestones in this field. Allee (1931), Shlaifer (1938); Geyer and Mann (1939), Job (1955); Delco (1963); Uematsu (1971); Parker (1973), Itazawa et al, (1978); Smatresk and Herreid (1980); Kanda and Itazawa (1981); Singh and Munshi (1985) and Jha (1995) have observed that grouping lowers down the oxygen uptake of fishes but the reverse relationship was also reported by Bowen (1932); Schuett (1934), Brett and Sutherland (1965), Umezawa and Kanikawa (1980), Ghosh et al, (1994), Krause and Ruxton (2002), Pandit and Ghosh (2007), Ward and Webster (2016). In due course, different aspects of grouping have been studied by various workers and many terms are proposed accordingly to explain the phenomenon.

In biology, any group of fish that stay together for social purposes is shoaling, and if the group is swimming in the same direction in a very organized manner, they are schooling. In common usage, the terms are sometimes used rather loosely. When fish, shrimp or other aquatic creatures swim together in an exceedingly loose cluster, this can be typically called a shoal. It is a combination of different species. A school is a group of the same fish species swimming together in synchrony, turning, twisting and forming sweeping, glinting shapes in the water.

An aggregation of fish is that the general term for any collection of fish that has been collected in some locality. The aggregations of fish can be structured or unstructured. An unstructured aggregation could well be a group of mixed species and sizes that have been randomly collected in the vicinity of some local resource, such as food or nesting sites.

If additionally, the aggregation comes together in an interaction socially, it will be said to be shoaling. Although shoaling fish could relate to every other in a very loose way, with each fish swimming as well as foraging somewhat independently, they are nonetheless attentive to the other members of the group as shown by the way they adjust behaviour like swimming, to remain near the other fish in the group. Shoaling groups can include fish of disparate sizes and may include mixed-species subgroups.

If the shoal becomes more closely coordinated, with the fish synchronising their swimming so that they all travel at the same speed and in the same direction, then the fish may also be assumed to be schooling (Helfman et al, 1997). Schooling fish are typically of the same species and thus the same age/size. Fish schools move with the individual members precisely apart from each another. The schools indulge in complex manoeuvres as if the schools had their own minds (Moyle and Cech, 2003).

Shlaifer (1938) observed the effect of group size in goldfish and explained that the soothing reduces the excitability of fish and consequently the metabolic rate. The soothing effect or lowering of oxygen uptake in fish was described as a calming effect by Parker (1973). Group effect in terms of either soothing or excitatory effect was discussed by Umezawa et al, (1983). The investigation of Kennedy and Pitcher (1975) has provided the concept of shoaling behaviour in fishes. The shoaling may be defined as a group of fishes for some social reasons. The fish become restless and excitable when away from the association of isomorphic intraspecific members. The shoaling fish during swimming get almost bounded to school on the way. Thus, the schooling may be defined as synchronized and polarized swimming of fishes in a shoal. The schooling effect to vision has been worked out by Malyukina et al, (1962) in a minnow (*Phloxinus phloxinus*) and goldfish (*Carassius cararssius*). Grigor'eva (1966) studied the effect of schooling on the respiratory metabolism of fishes.

According to Kanda and Itazawa (1986), the group effect is known in almost all freshwater fishes. However, he has dealt with this in a limited number of species. Group effect has a profound effect on the behaviour of individual fish. Fish in a small group are more timid, nervous and consequently have higher respiratory rates (ltazawa et al, 1978). This is because the fishes in small shoals are likely to detect a predator early and are less effective in escaping from an attack; the nervous fish shows a higher respiratory rate.

Despite the massive work done throughout the past two decades in terms of understanding the costs and edges of schooling and therefore the rules underlying the behaviour of fishes as part of a school, still, as of schools as a whole (Krause and Ruxton 2002; Ward and Webster 2016), little or no information is known regarding the effect of environmental (abiotic) factors on schooling behaviour.

Studies comparing the oxygen consumption of single fish versus that of fish during a group have shown that the calculated oxygen consumption of a group composed of individuals is lower than the sum of single fish. This may, however, also be associated with the calming effect of being in a group (Parker 1973; Nadler et al, 2016); therefore reliable estimates of the energetic advantage of swimming in a school need to take into consideration kinematic analyses in addition to metabolic measurements. An investigation by Abrahams and Colgan (1985) circumvented the issue of the confusing, calming effect by constrasting a school of fish with single fish isolated from their neighbours with simple partitions (Abrahams and Colgan, 1985). They found that fish in a school saved about 13% compared with the total of fish swimming on an individual basis.

Some gregarious species such as grey mullets boarding coastal areas, lagoons and estuaries; therefore they will be subject to hypoxic conditions seasonally. In addition, pelagic species living in large schools such as herring have been recorded in hypoxic areas like the Kattegat and a few Norwegian fjords (Domenici et al, 2002). According to Dommasnes et al, (1994), large schools of herring may have been the key cause of hypoxia in wintering areas.

Among the Indian fishes, oxygen uptake to group size was determined by Singh and Munshi (1985) in a goby (*Glossogobius giuris*). Ghosh et al, (1994) reported a significant increase in aquatic oxygen uptake in the juveniles of *Heteropneustes (=Saccobranchus) fossilis* in a group. Jha (1995) has worked out oxygen uptake during the early developmental stages of rohu (*Labeo rohita*) and catla (*Catla catla*) with special reference to grouping. Such experiments with Indian air-breathing fishes are fragmentary (Ghosh et al, 1994). Therefore, an attempt was made to study the effect of group size on oxygen uptake during the early developmental stages of two air-breathing fishes namely *Anabas testudineus* and *Channa punctatus*.

The results of the present work may be used for the transportation of different stages of live fish because during transportation the fishes are usually packed together in a group.

Materials and Methods

Proper sanitization of all the equipment and accessories is one of the essential steps for the determination of oxygen uptake of fishes under association amid the pandemic of COVID-19.

Isomorphic juveniles of *Anabas testudineus* and *Channa punctatus* were segregated from a large plastic pool and cemented tank respectively in which fishes were artificially bred by injecting heteroplastic pituitary extract. The juveniles were brought to the laboratory about a week before the experimentation and were kept separately in a plexiglass aquarium (60×30×30 cm). Fishes were fed with trash fishes and goat liver on an alternate day during acclimatization but feeding was sopped one day before experimentation.

A continuous water flow system of cylindrical glass respirometer (Ghosh and Munshi, 1987) was used for this experiment. The respirometer was connected to a water reservoir (30cm) maintained with a constant hydrostatic pressure through which uniform water flow was maintained through the respirometer. The juveniles were taken in the respirometer and were further acclimatized for about 2 hours to avoid the effect of handling. Care was taken so that each juvenile is given more or less the same space for its movement. To ensure this, respirometer of different sizes were used according to body weight and the number of fishes taken. A black cloth was used to cover the respirometer so that the juveniles were not exposed to external disturbances. After the first reading, subsequent readings were taken at an interval of 30 minutes. Two water samples were collected from two ends of the respirometer, one collected before the water enters the respirometer (inspired) and the other from the end where water comes out (expired) of the respirometer.

The concentration of dissolved oxygen in inspired and expired water was determined by Winkler's volumetric method (Welch, 1948). The difference between the amount of dissolved oxygen in inspired and expired water, together with the rate of water flow per unit time through the respirometer and bodyweight of the juveniles was used to calculate the rate of oxygen uptake per unit time and per unit body weight. First of all, oxygen uptake of a solitary juvenile was determined and then it was measured for juveniles in a group of gradually increasing numbers.

The variation in oxygen uptake due to grouping was determined under the following conditions:

(a) **In continuous flowing water:** It was worked out with the juveniles of *Anabas testudineus* (body weight: 0.6 to 1.0g and total length: 35 to 40 mm) and *Channa punctatus* (body weight: 0.7 to 1.0g and total length: 40 to 46 mm) at 21.5 ±1.0°C by using a cylindrical glass respirometer of 0.07 to 0.40 dm^3 size with a space of 0.306 ± 0.176 $dm^3 . g^{-1}$. The number of juveniles was gradually increased from 5, 10, 15 up to 40. The rate of water flow was gradually increased with an increase in the number of juveniles so that they feel comfortable. The variation in oxygen uptake due to group effect was calculated by the following formula as suggested by Itazawa et al, (1978).

$$\text{Group Effect} = \frac{\dot{V}O_2\text{in a smaller group} - \dot{V}O_2\text{in a larger group}}{\dot{V}O_2\text{in a smaller group}} \times 100$$

(b) **In still water:** Oxygen uptake due to group effect in still water was also determined to compare the values with that of continuously water flowing condition. Juveniles of body weight 4.50 to 5.00g of *Anabas testudineus* and 1.00 to 1.60g of *Channa punctatus* were used for this experiment. In this work, oxygen uptake due to group effect was measured at 30.0±1.0°C both in controlled as well as experimental water. The number of fishes was increased at an interval of 5 up to 30. For the determination of oxygen uptake from still water, water flow was prevented for 30 minutes through the respirometer. Per cent variation in oxygen uptake in experimental condition as compared to controlled one was calculated using the following formula:

$$\text{Group Effect} = \frac{\dot{V}O_2\text{in a controlled fish}(\text{es}) - \dot{V}O_2\text{in experimental fish}(\text{es})}{VO_2\text{in a controlled fish}(\text{es})} \times 100$$

The regression analysis using logarithmic transformations was made GraphPad Prism 5.04 to establish the relationship between oxygen uptake and group size under different conditions.

Observations and Results

The average aquatic oxygen uptake of the isomorphic juveniles of *Anabas testudineus* and *Channa punctatus* with group size under continuously flowing water system has been summarized in Tables-13 and 14. The effect of group size on the aquatic oxygen uptake of both the fishes under controlled (=continuously flowing water) and experimental (= still water) conditions have been given in Tables-15 and 16. The relationship between group size and aquatic oxygen uptake under different conditions has been presented in Table-17.

Behaviour of *Anabas testudineus* and *Channa punctatus*: To study the behaviour under laboratory conditions, 15 juveniles of *Anabas testudineus* and *Channa punctatus* were kept separately in a plastic pool having tap water. The juveniles usually swim separately and randomly but they prefer to remain in a shoal or association, sometimes only in a small group (=shoal) of only 2 to 4 juveniles. The juveniles have been found to prefer the corners or margins of the containers. The juveniles of *Anabas testudineus* keep a certain distance within a group.

Effect of group size on the aquatic oxygen uptake in flowing water

In *Anabas testudineus*: A solitary juvenile of *Anabas testudineus* (body weight: 0.92±0.19g) consumes 0.192±0.030 ml $O_2.g^{-1}.h^{-1}$. With a successive increase in group size 5 to 40 fishes at an interval of 5 fish each, the weight-specific oxygen uptake decreased from 0.140±0.026 to 0.080±0.006 $mlO_2.g^{-1}.h^{-1}$ (Table-13). Thus, with an increase in group size from 5 to 40 fish, the weight-specific oxygen uptake of an individual *Anabas testudineus* in a group decreased from 27 to 58% in comparison to a solitary fish. The reduction in oxygen uptake as compared to the solitary fish was found to be significant when the fishes were in a group of 20 onwards. A negative correlation ($r= -0.8411$, $p<0.01$) was obtained between aquatic oxygen uptake of a fish and the group size (i.e. the number of fish in a group) with a slope of -0.2647 (Table-17).

In *Channa punctatus*: A solitary juvenile of *Channa punctatus* of body weight 0.64 ± 0.03g consumes 0.505±0.088 $mlO_2.g^{-1}.h^{-1}$ under isolation. With a gradual increase in the number of juveniles from 5 to 40, the weight-specific oxygen uptake gradually decreased from 0.361±0.104 to 0.174±0.083$mlO_2.g^{-1}.h^{-1}$ (Table-14). Compared to the solitary fish, there is a gradual increase in the percent decrease of oxygen uptake per unit body weight from 29 to 66 %. A highly significant but negative correlation ($r=-0.9619$; $p<0.001$) was found between group size and oxygen uptake per unit body weight with a slope of -0.2798 (Table-17). In comparison to the solitary fish, the decrease in weight-specific oxygen uptake was found to be significant, when the juveniles were in a group of 5 to 40.

Effect of group size on aquatic oxygen uptake in still water

In this set of experiments continuous water-flowing condition was considered as control while the still water was taken as an experimental one. Under the controlled condition, juveniles of both *Anabas testudineus* as well as *Channa punctatus,* showed a decrease in weight-specific oxygen uptake due to group size. In experimental condition also, both the fishes gave a similar trend in weight-specific oxygen uptake or reduction due to group size.

In *Anabas testudineus*: The juveniles of *Anabas testudineus* consumed more oxygen in still water as compared to the fishes in continuously flowing water (Table-15). With the gradual increase in group size from 5 to 30 at an interval of 5 fish, the weight-specific oxygen uptake decreased from 0.053±0.070 to 0.012±0.003 $mlO_2.g^{-1}.h^{-1}$ in an experimental condition. The oxygen uptake in still water as compared to the oxygen uptake of a fish in continuously flowing water was found to increase. However, a decreasing trend from 165 to 50 % with the increase in group size. The variation in weight-specific oxygen uptake in still water as compared to continuous flowing water was found to be statistically highly significant ($r= -0.9845, p<0.001$) (Table-17). The reduction in oxygen uptake in experimental conditions as compared to the solitary fish was found to be significant when the fishes were in a group of 5 to 30.

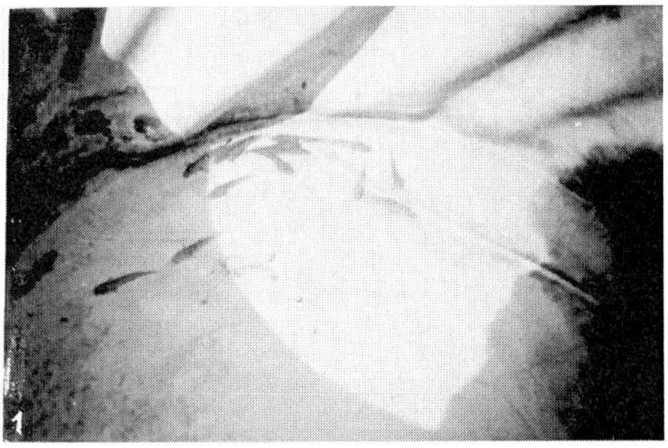

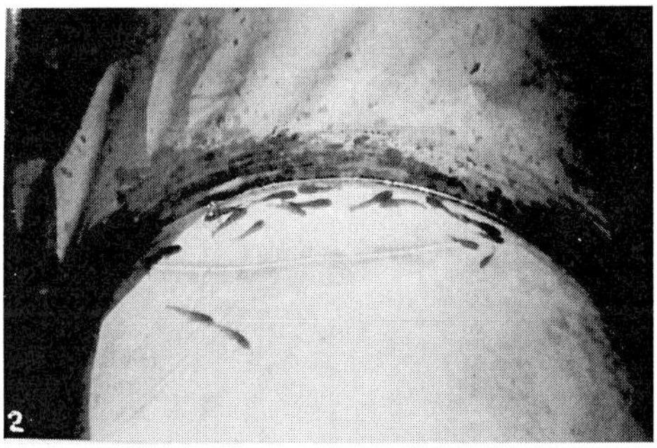

1. Juveniles of *Channa punctatus* showing shoaling behaviour
2. Juveniles of *Anabas testudineus* showing shoaling behaviour

Table 13: Effect of group size on the aquatic oxygen uptake in the early development stages of *Anabas testudineus* at 21.5±1.0°C.

S. No.	No. of fishes	Weight of a fish (g)	Aquatic Oxygen uptake		
			$mlO_2.g^{-1}.h^{-1}$	95%CL	% decrease due to GE
1.	01	0.92±0.19	0.192±0.030 (0.2178)	0.1094 0.4146	-
2.	05	0.90±0.12	1.40±0.026* (0.2190)	0.0798 0.2426	27.08
3.	10	0.74±0.06	0.150±0.018 (0.2307)	0.0677 0.1981	21.80
4.	15	0.98±0.10	0.127±0.004 (0.2141)	0.0609 0.1776	33.85
5.	20	1.04±0.07	0.098±0.018* (0.2108)	0.0563 0.1651	48.96
6.	25	1.18±0.08	0.058±0.008** (0.2039)	0.0528 0.1563	69.79
7.	30	1.05±0.16	0.089±0.008* (0.2103)	0.0500 0.1498	53.64
8.	35	1.12±0.10	0.087±0.007* (0.2067)	0.0478 0.1446	54.68
9.	40	0.90±0.12	0.080±0.006* (0.2190)	0.0459 0.1404	58.39

(*=p<0.05, ** = p<0.01) (Data in parenthesis = Estimated value) (CL = Confidence limit, GE = Group Effect).

Table 14: Effect of group size on the aquatic oxygen uptake in the early development stages of *Channa punctatus* at 21.5 ± 1.0⁰C.

S. No.	No. of fishes	Weight of a fish (g)	Aquatic Oxygen uptake		
			$mlO_2.g^{-1}.h^{-1}$	95%CL	% decrease due to GE
1.	01	0.64±0.03	0.505±0.088 (0.6185)	0.3996 0.7443	-
2.	05	0.74±0.05	0.361±0.104* (0.5935)	0.2681 0.4507	28.55
3.	10	0.72±0.01	0.290±0.076* (0.5972)	0.2228 0.3679	42.59
4.	15	0.77±0.04	0.277±0.001* (0.5861)	0.1991 0.3282	45.19
5.	20	0.78±0.04	0.274±0.031* (0.5851)	0.1834 0.3032	45.80
6.	25	0.89±0.04	0.236±0.001* (0.5634)	0.1720 0.2855	53.33
7.	30	0.99±0.01	0.205±0.127* (0.5468)	0.1630 0.2720	59.45

S. No.	No. of fishes	Weight of a fish (g)	Aquatic Oxygen uptake		
			$mlO_2.g^{-1}.h^{-1}$	95%CL	% decrease due to GE
8.	35	1.01±0.12	0.177±0.133* (0.5438)	0.1557 0.2612	64.87
9.	40	1.02±0.02	0.174±0.083** (0.5431)	0.1496 0.2523	65.52

(*=p<0.05, ** = p<0.01) (Data in parenthesis = Estimated value) (CL = Confidence limit, GE = Group Effect).

In ***Channa punctatus*****:** During the experimental condition, with the gradual increase in group size (5 to 30 with a unit increase of 5 fish), the weight-specific oxygen uptake in still water decreased from 0.820 ± 0.180 to 0.072 ± 0.030 $mlO_2.g^{-1}.h^{-1}$. In this fish also, the juveniles consumed more oxygen in still water when compared to the fishes in continuously following water (Table-16). The decrease in oxygen uptake during the experimental condition of a fish in a group when compared to the solitary fish was highly significant (r=0.9923, p <0.001) with a slope -0.6774 (Table-17). However, no definite trend was observed in percent increase in oxygen uptake during experimental conditions than a controlled one. The reduction in oxygen uptake as compared to a solitary fish under both the controlled and experimental conditions was found to be statistically significant when the fishes were in a group of 5 to 30. The changes in oxygen uptake under continuously flowing and still water conditions were also found to be significant.

Table 15: Effect of group size on the aquatic oxygen uptake in the early developmental stages of *Anabas testudineus* under continuous flowing (=control) and still (-experimental) water condition at 30.0 ±1.00 °C.

Sl.No.	No. of fishes	Weight of a fish (g)	Aquatic oxygen uptake ($mlO_2.g^{-1}.h^{-1}$)					
			Continuous flowing water		Still water observed	% decrease due to GE	% increase in VO_2 in still water in comparison to continuous flowing water	Changes in oxygen uptake under continuous flowing and still water conditions
			Observed	95%CL				
1	01	4.75±0.35	0.091±0.040 (0.0310)	0.0023 1.3486	0.213±0.067 (0.0645)		134.00	$p<0.05$
2	05	5.45±0.17	0.020±0.011* (0.0295)	0.0022 0.4172	0.053±0.010* (0.0579)	78.02	165.00	$p<0.05$
3	10	5.32±0.18	0.018±0.008* (0.0297)	0.0018 0.3007	0.045±0.008* (0.0590)	80.22	150.00	$p<0.05$
4	15	5.32±0.18	0.013±0.007* (0.0297)	0.0015 0.2632	0.031±0.007* (0.0590)	85.71	134.46	$p>0.05$
5	20	5.46±0.02	0.011±0.004* (0.0294)	0.0013 0.2457	0.022±0.002* (0.0578)	87.91	100.00	$p>0.05$
6	25	5.22±0.16	0.009±0.004* (0.0300)	0.0012 0.2362	0.017±0.006* (0.0599)	90.11	88.88	$p<0.05$
7	30	5.25±0.09	0.008±0.002 (0.0299)	0.0010 0.2307	0.012±0.003* (0.0596)	91.21	50.00	$p<0.05$

(*=$p<0.05$, ** = $p<0.01$) (Data in parenthesis = Estimated value) (CL = Confidence limit, GE = Group Effect).

Table 16: Effect of group size on the aquatic oxygen uptake in the early developmental stages of *Channa punctatus* under continuous flowing (=control) and still (-experimental) water condition at 30.0±1.0⁰ C.

Sl.No.	No. of fishes	Weight of a fish (g)	Aquatic oxygen uptake ($mlO_2.g^{-1}.h^{-1}$)					
			Continuous flowing water		Still water observed	% decrease due to GE	% increase in VO_2 in still water in comparison to continuous flowing water	Changes in oxygen uptake under continuous flowing and still water conditions
			Observed	95%CL				
1	01	1.03±0.30	0.534±0.130 (0.5725)	0.2117 1.6359	0.820±0.180* (0.9480)		54.72	$p<0.05$
2	05	0.79±0.08	0.164±0.050* (0.6930)	0.0796 0.4280	0.303±0.060* (0.3187)	69.29	87.50	$p<0.05$
3	10	0.81±0.03	0.148±0.060* (0.6806)	0.0494 0.2544	0.296±0.010* (0.1993)	72.28	107.14	$p<0.05$
4	15	0.69±0.09	0.130±0.040* (0.7639)	0.0366 0.1912	0.232±0.060* (0.1514)	75.65	78.46	$P<0.01$
5	20	1.39±0.14	0.080±0.010* (0.4613)	0.0294 0.1574	0.186±0.010* (0.1246)	85.02	132.50	$p>0.05$
6	25	1.19±0.00	0.059±0.020* (0.5159)	0.0247 0.1360	0.109±0.000* (0.1071)	88.95	84.75	$p<0.01$
7	30	1.38±0.36	0.036±0.010* (0.4637)	0.0213 0.1210	0.072±0.030* (0.0947)	0.0353 0.2539	100.00	$p>0.05$

(*=p<0.05, ** = p<0.01) (Data in parenthesis = Estimated value) (CL = Confidence limit, GE = Group Effect).

Table 17: Intercept (=a), slope (=b) along with their 95% confidence limits and standard error and correlation coefficient of *Anabas testudineus* and *Channa punctatus* to show relation between oxygen uptake and different parameters.

Fish	Group size v/s oxygen uptake ($mlO_2.g^{-1}.h^{-1}$) under	Intercept (=log a)			Slope (=b)			Correlation coefficient
		Estimated	95% CL	SE	Estimated	95% CL	SE	
Anabas testudineus	a. Continuous water flow	-0.6716	-0.8590 -0.4842	0.0792	-0.2647	-0.4169 -0.1126	0.0643	0.8411 ($p<0.01$)
	b. Controlled (continuously flowing) water	-1.2176	-2.1669 -0.3381	0.3557	-0.6834	-2.1562 0.7965	0.5744	0.9860 ($p<0.001$)
	c. Still water	-0.6587	-	0.0692	-0.7857	-	0.0621	0.9847 ($p<0.001$)
	d. Controlled water v/s still water	-0.9315	-	0.6892	0.2914	-	0.4107	0.9845 ($p<0.001$)
Channa punctatus	a. Continuous water flow	-0.2633	-0.3508 -0.1758	-	-0.2798	-0.3509 -0.2088	-	0.9619 ($p<0.001$)
	b. Controlled (continuously flowing) water	-0.2303	-0.5240 -0.0634	0.1143	-0.7202	-0.9837 -0.4567	0.1025	0.9529 ($p<0.001$)
	c. Still water	-0.0232	-	0.1299	-0.6774	-	0.1265	0.9333 ($p<0.001$)
	d. Controlled water v/s still water	0.2052	-	0.0543	0.9529	-	0.0531	0.9923 ($p<0.001$)

Discussion

Determination of the amount of oxygen uptake by a fish is an important problem in the studies where the effect of grouping on respiration is investigated. The variation in oxygen uptake in fishes due to grouping or association is probably related to their behaviour. Fishes generally show either soothing (=calming) behaviour (Umezawa et al, 1983), excitatory (=aggressive) behaviour (Umezawa et al, 1983), shoaling behaviour (Kennedy and Pitcher, 1975) or schooling behaviour (Grigor'eva, 1966) under grouping. Shoaling is a type of social behaviour possessing the importance of individual differences and regarded as analogous to flocking in birds. It is based on the rule governing the decision of individual fish to join, stay with or leave the group but no implication for structure and function. The shoaling fish enhances counter predator attack by avoidance, diluting the probability of capture of a predator behaves homogeneously, which could increase the accuracy of homing and foraging benefits. The shoaling fishes during swimming get almost bounded to a school on the way. Shaw (1970) explained that a school has no apparent behavioural differentiation among members, no leaders, no persistent ties, no pair bonds and no dominance relationship. Schooling fishes are of similar size, move at the same and uniform speed tending to perform the same activity together. The school assesses the costs and benefits to individuals but is restricted to turning and accelerating fish towards a pole in a shoal. It means that schooling is a type of shoaling behaviour under certain conditions and is often seen among small fishes in shallow water. These fishes become restless and excited when away from the association of intraspecific isomorphic members.

In the present experiment, it has been observed that oxygen uptake in the early developmental stages of *Anabas testudineus* and *Channa punctatus* was less in grouped fishes or a fish in a larger group than in an isolated fish or a fish in a smaller group. The oxygen uptake of these fishes under different conditions of grouping was compared in terms of per unit body weight (mlO_2.-g^{-1}.h^{-1}) because the size of fishes used for experimental purposes was more or less the same with very limited variation in their body weight.

The weight-specific oxygen uptake in *Anabas testudineus* and *Channa punctatus* under continuously flowing water decreased both with the successive increase in the number of fish from 5 to 40 at an interval of 5 fish. The reduction in oxygen uptake is statistically significant. The average reduction in oxygen uptake was calculated to be 46 ± 16 % in *Anabas testudineus* while 51±12% in *Channa punctatus* (Tables- 13 and 14). Itazawa et al, (1978) reported a 14.97 ± 1.75% reduction in weight-specific oxygen uptake due to grouping while working with medaka or Japanese killifish (*Oryzias latipes*) of body weight 0.484 ± 0.027g. Itazawa and Watanabe (1979) also investigated the effect of grouping on respiration of medaka (body weight: 120 to 400mg) and found that mean oxygen uptake of a fish was more in isolation than a fish in a group of 10. In *Catla catla*, Jha (1995) calculated 35.66 ± 23.88% decrease in weight-specific oxygen uptake with an increase in group size from 1 to 30. The result of the present study is in line with the above-referred works. However, Bowen (1932) found no significant difference in oxygen uptake between grouped and isolated catfish (*Ameriurus melas*) although an isolated fish exhibits a higher rate of oxygen uptake. In fathead minnow, *Pimephales promelas* (body weight: 0.47 to 2.33g), a reduction in oxygen uptake occurs in a group of 2 and 3 fishes only and a further increase in group size, increases the oxygen uptake (Wares II and lgram. 1979). Imabayashi and Takahashi (1987) studied the changes in oxygen uptake of larvae and juveniles (body weight: 3.75 – 74.6 mg) of red seabream, *Pagrus major* with group size and found that oxygen uptake of a solitary fish was 9.3% higher than that of fishes in a group of 2 but 19.2% lower than that of fishes in a group of 8. They suggested that group effect on oxygen uptake is correlated with the change of behaviour with growth and development.

In the present study, both in *Anabas testudineus* and *Channa punctatus*, the reduction in weight-specific oxygen uptake by a fish in a group of 20 and onwards was found to be statistically significant when compared with a solitary fish or a fish in isolation. The graphical representation indicates that oxygen uptake in an *Anabas testudineus* sharply decreases in a group of 5 and then a gradual reduction occurs in a group of 10 to 40 fishes. However, the weight-specific oxygen uptake in a *Channa punctatus* shows sharp reduction up to a group of 10 fish. The above results suggest that the isomorphic juveniles of *Anabas testudineus* and *Channa punctatus* probably feel easy in a group due to their companionship.

In the second experiment, it was seen that oxygen uptake was always lower in fishes under grouping than in isolation both in controlled (=continuously flowing water) and in experimental (=still water) conditions. Under the experimental conditions, the decrease was gradual (134 to 50 %) with an increase in group size. The rate of weight-specific oxygen uptake of a juvenile of *Anabas testudineus* decreases by a power of -0.7587 when the number of fishes was increased up to 30. Similarly, the oxygen uptake of *Channa punctatus* in experimental conditions got decreased by a functional power of -0.6774 with an increase in group size (Table-17). In *Anabas testudineus*, the increase in oxygen uptake was 115±39 % when compared with the controlled conditions, while the increase was 80 ± 20 % in *Channa punctatus*. The rate of oxygen uptake in experimental conditions was however always more than that of a fish under controlled conditions. When percent increase in oxygen uptake under experimental conditions was considered, a decreasing trend was seen in *Anabas testudineus* but no definite trend in *Channa punctatus*.

A juvenile of *Anabas testudineus* and *Channa punctatus* shows a sharp reduction in both the controlled and experimental conditions when they are in a group of 5. However, a slight variation was observed in oxygen uptake by a juvenile in a group of 10 than when they are in a group of 5. There is a gradual decrease in oxygen uptake of a fish when they are in a group of 15 fishes and onwards. But, the rate of decrease was more in *Channa punctatus* as compared to *Anabas testudineus*. Jha (1995) also reported 31.78 ± 14.24% increase in oxygen uptake in *Catla catla* (body weight: 1.48 ± 0.106g) under the experimental conditions in comparison to the controlled conditions. Therefore, the present work corroborates with the work of Jha (1995). These results show that both *Anabas testudineus* and *Channa punctatus* follow similar trends in their nature of oxygen uptake. The higher rate of oxygen uptake in the experimental condition was probably due to the stressful condition of fish, a decrease in the amount of dissolved oxygen in water, etc.

It seems that the oxygen uptake of a fish is regulated by their behaviour which mutually contributes to the total oxygen uptake. Since both the experimental fishes (*Anabas testudineus* and *Channa punctatus*) are predatory, it is expected that the oxygen uptake will increase due to grouping, as reported by Brett and Sutherland (1965), Wirtz and Davenport (1976), Umezawa et al, (1981) and Ghosh et al, (1994), respectively in pumpkinseed, blennies, goby and *Heteropneustes fossilis*. As far as the behaviour of *Anabas testudineus* is concerned, it did not show schooling behaviour during their early developmental stages. However, their juveniles were found to behave homogeneously to form a small shoal of 2-4 individuals. The reduction in oxygen uptake in an *Anabas testudineus* in the grouped condition is probably a result of such shoaling behaviour which creates a soothing or calming effect. The early developmental stages of *Channa punctatus* although show schooling behaviour till they attain a length of about 30 mm (body weight: 230 mg). But, as they grow in size the schooling behaviour is gradually lost. The juveniles still have been found to take the corner of an aquarium/plastic pool and they prefer to remain in association with a group of 3-5 fishes. They seem to feel at ease in a shoal. Macdonald also described that *Channa punctatus* feed in a shoal. Therefore, it may be summarized that the decrease in oxygen uptake in this fish is also probably due to the soothing effect in a shoal.

5

Effect of Anaesthetic on Oxygen Uptake

Introduction

In fish culture, anesthetic agents are commonly used in vaccination, labeling, and various measurements (Gomes et al, 2006; Collymore et al, 2014). But no agents have been able to obtain long-duration anesthesia in fishes. The use of conventional anesthetic agents in fisheries is not beneficial in terms of environmental safety as the discarded solutions directly reach oceans and rivers. The use of anesthetic agents presenting a risk of staying in the tissues of cultivated fish has been warned against. For example, the Japanese Ministry of Agriculture, Forestry, and Fisheries caution against the use of anesthetic agents within 7 days before fishes and shellfishes are caught.

The use of anaesthetics has gained considerable importance in Fishery Science particularly during spawning (Colesante, 1977; Kidd and Banks, 1990) and transportation of live fishes or their early developmental stages (Maheshwari 1990; Jhingran, 1991; Pandit and Ghosh, 2012). This is because handling causes stress and physical damage to the fish, loss of mucilaginous covering, which in turn makes the fish susceptible to attack by bacteria and fungi (Brown, 1987). The anaesthetics are soothing medicines that reduce the metabolism and control the excitability as well as stress in a fish. Recently, Kugino et al, (2016) investigated whether the addition of nano oxygen-carrying bubbles to water at the same time as dissolved carbon-dioxide could result in safe, long-duration fish anesthesia.

Important features considerations to be considered for the selection of anaesthetics include its effectiveness, cost, availability, ease of use, safety aspects and ultimately the purpose of the experiment and fish species used (Mylonas et al, 2005). Many safe and food-grade anaesthetics are used for this purpose, which causes temporary loss of sensation. The proper dosage is critical and in general, it varies widely with the type of anaesthetics, species employed as well as size and condition of the fish.

Usually, the brooders are sedated either by dissolving the anaesthetics in the water, directly injecting the drug into the fish body (Kewalramani and Gagote,

1968) or spray of anaesthetics on the fish body (Colesante 1977; Kidd and Banks, 1990). The application of anaesthetics reduces the oxygen uptake and also the release of carbon dioxide along with other excretory products. In India, the use of anaesthesia in Fisheries, mainly for transportation and tagging, dates back to the early 1960s. Smith and Bell (1967) have discussed the properties of anaesthetics. The techniques of application of anaesthetics were described by Klontz and Smith (1968), whereas, stages of anesthesia and hazards of its use were explained by Mc Farland and Klontz (1969).

Some 30 compounds of the commonly used fish anaesthetics are Sodium amytal, Novocaine, Chloral hydrate, Amobarbital sodium, Barbital sodium, Chloretone, Methyl pentynol, Quinaldine (2-methylquinoline), Hydroxy quinaldine, Propaxate (R7464), Methyl paraphynol (Dormison), Urethane (Ethyl carbamate), t-Amyl alcohol, Amylene hydrate, Thiouracil, MS-222 (Ethyl-m-aminobenzoate), Metomedate, AQUI-S, Eugenol and Benzocaine (Ethyl-p-aminobenzoate) (Chen, 2019). Besides these hypothermia (=cooling of fishes to about 4°C) is one of the oldest methods for anaesthetizing the fish. Sodium amytal is a soft water anaesthetic showing antagonism with calcium and mostly tried to transport live fishes. The use of quinaldine at 5 ppm successfully transports brood fish for 5 hours without any mortality (Natarajan and Ranganathan, 1960). Novocaine was regularly used to sedate major carps in transportation by the Department of Fisheries, Bombay since 1966. Jhingran and Pullin (1988) stated that anesthetization of brood fish is essential before transportation which is not necessary for fingerlings.

In India, Kewalramani and Gogate (1968) have used different anaesthetics such as Novocaine, Amobarbital sodium and Barbital sodium to sedate certain Indian major carps. Sharma et al, (1991) reported that MS-222 is effective in narcotizing *Cirrhinus reba* and *Barbus mahicola* at 50 ppm. Basavaruju et al, (1998) studied that *Catla* is more delicate than common craps and require lower doses of Benzocaine for effective anaesthetisation. Schoettger and Julin (1967) have studied the response of MS-222 in salmonids.

Sylvester and Holland (1982) had investigated the effect of water temperature, hardness and stocking density on MS-222 response in rainbow trout (*Salmo gairdneri*), common carp (*Cyprinus carpio*) and fathead minnow (*Pimephales promelas*). Novocaine is generally used in surgical operations and relief of pain. Barbiturates are usually given for alleviating mild discomforts, etc. Phenobarbital and Barbital sodium are long-acting drugs. Pentobarbital and thiomyl (derivative of Barbiturates) induce anaesthesia for short or very short durations. Of all the above, MS-222 is a registered fish anaesthetic in the United States (Meyer et al, 1976).

In Poland, yet another anaesthetic was used, Propanidid, which was considered to be a safe and very effective anaesthetic for use with fish, but is no longer being produced (Siwicki 1984, Veenstra et al. 1987). Consequently, new less toxic and more effective preparations are required for inducing general anaesthesia in fish. A new drug, Propiscin, was developed at the Inland Fisheries Institute in Poland. The active ingredient is etomidate, which is commonly used to anaesthetize fish (Morgan et al. 1975, Vermeer and Falls 1988).

Propiscin has proved to be very successful and can induce a short period of general anaesthesia which lasts about 30 min (Szkudlarek and Zakêœ 1996). It has all the advantages of general anaesthetic agents: it does not act depressively on the respiratory system; it causes a slight drop in blood pressure; it is safe and, possibly most important, it is not teratogenic or carcinogenic.

The anaesthetics change the physiological behaviour of fishes but the information is fragmentary regarding its effect on the respiratory physiology of fish (Wedemeyer, 1970; Housten et al, 1971). Availability of oxygen plays a vital role in the survival of live fishes during their transportation.

The oxygen uptake is influenced by several factors more importantly the state of physical activity, size and species of fish, etc. The oxygen uptake increases due to handling and hyperactivity of the fish under stress. The smaller fishes have more weight-specific oxygen uptake. All these, along with accumulated metabolic waste in water tend to reduce the level of dissolved oxygen in the transporting medium.

The introduction of anaesthetics in the fish body through hypodermic injection is satisfactory in sedating the brooders. But its application in early developmental stages is limited. Jha and Ghosh (1993) have studied the effect of Benzocaine on the rate of oxygen uptake in the juveniles of Indian major carps, *Labeo rohita*.

Literature regarding the effect of anaesthetics on oxygen uptake in Indian air-breathing fishes is probably lacking. Pandit and Ghosh (1999) evaluated the effect of Benzocaine on the aquatic oxygen uptake in the juveniles of *Heteropneustes fossilis*. Oxygen consumption was measured in spotty (*Notolabrus celidotus*) over the 24 hours following anaesthesia by Hill and Forster (2004).

Guo et al, (1995) carried out the effects of various dosages of 2- methyl phenoxyethanol. quinaldine sulphate, MS-222 and metomidate on the oxygen consumption rates of two groups of platyfish, at three temperatures.

The stress associated with live fish handling and transportation is a major challenge and minimizing this can successfully improve the survival rate and associated biochemical quality changes of the fish. Anaesthetics are known for reducing stress and related problems during live fish transportation. During live transportation, anaesthetics are widely employed before and during transport to slow the metabolism of the fish, thus reducing the rate of oxygen uptake, the release of carbon dioxide, and ammonia production.

Muscle movements in the respiratory branchia are suppressed when fishes are anesthetized and oxygen uptake decreases by diffusive transfer. These results in a shortage of oxygen compared to the organism's oxygen demand, which leads to sudden death, even if the creature is put in water saturated with dissolved oxygen.

Therefore, for long-duration anesthesia to be achieved either the organism's oxygen demand must be decreased or increased dissolved oxygen must be supplied via over saturation of the aquatic environment.

Harmon (2009) reports also suggested the importance of preserving adequate water quality during transport as it reduces physiological stress and indicated these parameters could differ based on life stage, health and previous holding conditions within a species.

There is no evidence on the long-term impact of anaesthetics on fish oxygen consumption. Longer-term monitoring of post-anaesthesia oxygen consumption would not only elucidate the changes in oxygen consumption changes during recovery, but would also provide an estimate of recovery times.

Therefore, the present work has been undertaken to determine the effect of Benzocaine on the aquatic oxygen uptake in the early developmental stages of snake-headed murrel, *Channa punctatus* (Bloch).

The results from this study will be useful to Fishery Scientists working with this commercially important cultivable fish species especially for their packing and transportation purposes either for food, artificial breeding or pisciculture practices.

However, more intense studies are needed to understand the dosage levels for long-term transportation. Similarly, to standardize their permissible concentrations in food fish, toxicological investigations on the use of essential oil are to be intensely explored.

Materials and Methods

Proper sanitization of all the equipment and accessories is one of the essential steps for the determination of oxygen uptake of fishes under sedation and anaesthetization in this pandemic of COVID-19.

Early developmental stages of *Channa punctatus* were collected from neighbouring rural areas and were kept in a large plastic pool and treated with methylene blue. After one week, the early developmental stages of 1.00 to 7.00g body weight were segregated and acclimatized in a plexiglass aquarium (60×30×30cm) having chlorine-free tap water for a week in the laboratory condition. During acclimatization, the early developmental stages were fed with trash fish and chopped goat liver on alternate days.

To determine the suitable dose of Benzocaine (Ethyl-p-aminobenzoate), five or selected number of fishes, a set of more or less similar weight were kept separately in smaller plexiglass aquaria (30×30×30 cm) in different doses (10 to 25 $mg.l^{-1}$) of unbuffered anaesthetic.

The reaction of fishes to the anaesthetic was observed during the first one hour followed by every half an hour up to a period of eight hours and then at an interval of 24 hours for 72 to 96 hours. Time to complete sedation (cessation of gill ventilation) was recorded for each fish. A dose of 18.0 ± 3.0 $mg.l^{-1}$ was found to be suitable for anesthetization of the early developmental stages of *Channa punctatus* (bodyweight=1.0 to 6.5 g and total length= 5.3 to 8.5cm). Fishes were then transferred to a benzocaine free fresh tap water for recovery. The recovered fishes were transferred to a plexiglass glass aquarium and were observed for 24 hours.

The rate of aquatic oxygen uptake by the early developmental stages of *Channa punctatus* in controlled and experimental conditions was determined by the use of the cylindrical glass respirometer with a continuous water flow system (Ghosh and Munshi, 1987) at 19.5±1.0° C. During experimentation, the respirometer was connected with a water reservoir having constant hydrostatic pressure. To avoid disturbance by visual factors, a black cloth was used to cover the respirometer.

Firstly, the oxygen uptake was determined under controlled conditions i.e., in chlorine-free tap water. The early developmental stages were exposed for 2h and 23h respectively in 18.0 ± 3.0 $mg.l^{-1}$ of Benzocaine before being introducing into the respirometer filled with anaesthetic water. The water reservoir was also filled with the same solution of Benzocaine. After an hour of acclimatization in the respirometer with a continuous flowing anaesthetic solution, oxygen uptake was determined by the difference in dissolved oxygen content of the inspired and expired anaesthetic water.

The amount of dissolved oxygen in the inspired and expired water was estimated by Winkler's volumetric method (Welch, 1948). The rate of aquatic oxygen uptake was calculated from the difference between the amount of dissolved oxygen in the inspired and expired water along with the rate of water flow per minute through the respirometer. Opercular frequency in controlled, as well as anaesthetic exposed conditions was counted by visual observation. Three readings of opercular frequency were taken at an interval of 5 minutes each during the experimentation.

Average values were calculated for all the observations. The statistical analysis (Student's t-test) was made with GraphPad Prism 5.04 software to compare the significance in the aquatic oxygen uptake and opercular frequency between control and anaesthetic exposed fishes.

Observations and Results

During experimentation, it was observed that the experimental specimens of *Channa punctatus* were less active in comparison to control fishes and settle down at the bottom of the respirometer having anaesthetic water with Benzocaine. The treated juveniles showed some loss of reactivity to external stimuli and reduction in opercular frequency. The data of oxygen uptake, opercular frequency and their respective calculated values are given in Tables-18 to 20.

Oxygen uptake per unit time: In controlled conditions, the rate of aquatic oxygen uptake per unit time in the early developmental stages of *Channa punctatus* increased from 0.183 ± 0.029 to 0.481±0.034 $mlO_2.h^{-1}$ with an increase in body weight from 1.45±0.18 to 6.12±0.11g. The 3h and 24 h Benzocaine exposed individuals registered an increment of oxygen uptake from 0.096 ± 0.015 to 0.286 ± 0.023 and 0.105±0.017 to 0.300±0.037 $mlO_2.h^{-1}$ respectively for similar body weight range. Under the controlled conditions, a fish of 3.71 ± 1.18 g average body weight consumes 0.332 ± 0.109 $mlO_2.h^{-1}$, whereas, in the 3h and 24h Benzocaine exposed fish, the average aquatic oxygen uptake was calculated to be 0.193±0.068 and 0.205±0.071 $mlO_2.h^{-1}$ respectively.

Thus, in comparison to control fish the 3h exposed individual shows 42.30±2.78% reduction in its aquatic oxygen uptake, while the reduction was 38.54±2.44% in the 24 h exposed fish. Application of t-test suggests that the reduction in aquatic oxygen uptake per unit time due to exposure of Benzocaine is highly significant at 3h ($r=0.9972$; $p<0.001$) and 24h ($r=0.9949$; $p<0.001$) exposure in comparison to the controlled fish (Table-18).

Oxygen uptake per unit body weight: Under the controlled condition, with the gradual increase in body weight (1.45±0.18 to 6.12+0.11 g) in *Channa punctatus*, the weight-specific aquatic oxygen uptake decreased from 0.127±0.020 to 0.0791 ± 0.006 $mlO_2.g^{-1}.h^{-1}$. However, in 3h and 24h Benzocaine exposed fish, aquatic oxygen uptake per unit body weight decreased from 0.066±0.010 to 0.047±0.004 and 0.073± 0.011 to 0.048±0.006 $mlO_2.g^{-1}.h^{-1}$ respectively. A fish of average body weight of 3.71±1.78 g consumes 0.097±0.019 $mlO_2.g^{-1}.h^{-1}$ under controlled conditions while the 3h and 24 h Benzocaine exposed individual of similar body weight consumed 0.056±0.009 and 0.059±0.010 $mlO_2.g^{-1}.h^{-1}$ respectively.

When compared to controlled conditions, the 3h Benzocaine exposed individual shows 42.43±3.03% reduction in its weight-specific aquatic oxygen uptake while 24h exposed fish shows only 39.14±2.09% reduction. Statistical analysis shows that in comparison to the controlled fish, the reduction in weight-specific aquatic oxygen uptake due to exposure of Benzocaine is highly significant ($p<0.001$) in 3 h (r=0.968) and 24 h (r=0.986) exposure (Table-19).

Opercular frequency: The opercular frequency of the control *Channa punctatus* (body weight: 1.45±0.18 to 6.12±0.12g) ranged between 46.0±1.0 to 49.0±1.0 but it ranged between 24.0±2.0 to 29.0±1.0 and 27.0±2.0 to 31.0±3.0 min^{-1} respectively in the individuals of 3h and 24h Benzocaine exposed fish of the similar body weight. The average opercular frequency 47.0±2.0 min^{-1} under the controlled conditions for a fish of body weight 3.71±1.78 g goes down to 27.0±2.0 min^{-1} in 3h and 29.0±2.0 in 24 h exposed individuals (Table-20). The opercular frequency, thus, showed about 42.81±3.21 % and 39.68 ± 2.95 % reduction in 3 h and 24 h Benzocaine exposed juveniles of *Channa punctatus*. Paired t-test analysis suggests that the reduction in opercular frequency due to Benzocane exposure at 3h (r=0.9943) and 24h (r=0.9524) is highly significant ($p<0.001$).

Table 18: Oxygen uptake $mlO_2.h^{-1}$ in controlled and experimental (18.0±3.0 ppm Benzocaine treated) *Channa punctatus* at 19.5 ±1.0° C

S. No.	Body weight of a fish (g)	Oxygen uptake ($mlO_2.h^{-1}$)				
		Control	3h Exposure		24 h Exposure	
				% reduction in comparison to control		% reduction in comparison to control
1.	1.45±0.18	0.183±0.029	0.096±0.015	47.54	0.105±0.017	42.62
2.	2.23±0.09	0.245±0.039	0.142±0.023	42.04	0.153±0.025	37.55
3.	3.15±0.17	0.308±0.085	0.179±0.050	41.88	0.194±0.032	37.01
4.	4.13±0.15	0.367±0.027	0.222±0.016	39.51	0.219±0.029	40.33
5.	5.18±0.26	0.407±0.023	0.235±0.013	42.26	0.260±0.047	36.12
6.	6.12±0.11	0.481±0.034	0.286±0.023	40.54	0.300±0.037	37.63
Average	**3.71±1.18**	**0.332±0.109**	**0.193±0.068**	**42.30±2.78*** **(41.87)**	**0.205±0.071**	**38.54±2.44*** **(38.25)**

(* = $p<0.001$)

Table 19: Oxygen uptake $mlO_2.g^{-1}.h^{-1}$ in controlled and experimental (18.0±3.0 ppm Benzocaine treated) *Channa punctatus* at 19.5 ±1.0° C.

S. No.	Body weight of a fish (g)	Oxygen uptake ()				
		Control	3h Exposure		24 h Exposure	
				% reduction in comparison to control		% reduction in comparison to control
1.	1.45±0.18	0.127±0.020	0.066±0.010	48.03	0.073±0.011	42.52
2.	2.23±0.09	0.110±0.018	0.064±0.010	41.82	0.068±0.011	38.18
3.	3.15±0.17	0.098±0.027	0.057±0.016	41.84	0.061±0.010	37.76
4.	4.13±0.15	0.089±0.006	0.054±0.005	39.32	0.053±0.007	40.45
5.	5.18±0.26	0.079±0.024	0.045±0.003	43.04	0.050±0.009	36.71
6.	6.12±0.11	0.079±0.006	0.047±0.004	40.50	0.048±0.006	39.24
Average	**3.71±1.18**	**0.097±0.019**	**0.056±0.009**	**42.43±3.03*** **(42.27)**	**0.059±0.010**	**39.14±2.09*** **(39.17)**

(* = p<0.001)

Table 20: Opercular frequency (min^{-1}) in controlled and experimental (18.0±3.0) ***Channa punctatus*** **at 19.5±1.0° C.**

S. No.	Body weight of a fish (g)	Opercular frequency (min^{-1})				
		Control	3h Exposure		24 h Exposure	
				% reduction in comparison to control		% reduction in comparison to control
1.	1.45±0.18	47.0±2.0	24.0±2.0	48.93	27.0±2.0	42.55
2.	2.23±0.09	48.0±2.0	28.0±2.0	41.67	30.0±2.0	37.50
3.	3.15±0.17	47.0±1.0	27.0±2.0	42.55	30.0±3.0	36.17
4.	4.13±0.15	48.0±3.0	29.0±1.0	39.58	28.0±1.0	41.66
5.	5.18±0.26	49.0±1.0	28.0±2.0	42.86	31.0±3.0	36.73
6.	6.12±0.11	46.0±1.0	27.0±3.0	41.30	28.0±2.0	43.48
Average	**3.71±1.18**	**47.0±2.0**	**27.0±2.0**	**42.81±3.21* (42.55)**	**29.0±2.0**	**39.68±2.95* (38.30)**

(* = p<0.001)

Discussion

Aquacultural research often requires handling of fishes for weighing, measurement, marking, tagging, breeding and collection etc., which may cause stress and physical damage to the fish. The use of anaesthesia makes the procedure easier, reduces the chances of physical damage and provides stress-free conditions to the fish. The anaesthetics are frequently used in large quantity with inadequate knowledge of long term effects either on its user or the fish itself.

According to Wood (1956), urethane (Ethyl carbamate) is readily absorbed through the skin but has both carcinogenic and leucopenic action, though it is widely used in tagging and marking salmonids. MS-222 (ethyl-m-aminobenzoate or tricaine methane sulphonate) is a widely accepted and registered fish anaesthetic in the United States (Meyer et al, 1976). However, MS-222 is not highly recommended for the commercial purpose of fish transportation probably due to its higher cost, toxic nature in seawater when exposed to light, some undesirable side effects, acidity including damage to the olfactory epithelium, hypoxia, hyperglycemia and other physiological changes (Bell, 1967; Ross and Ross, 1984 and Rodriguez-Gutierrez and Esquirvel-Herrera, 1995). Although, Locke (1969) has suggested that quinaldine (2-methyl quinoline) should replace MS-222, the fishes have been found to show some degree of responsiveness in quinaldine. So, it may be used along with MS-222 (Schoettger and Steucke, 1970). Methyl pentynol is a useful anaesthetic to reduce activity in salmonids (Svendsen, 1969) but a higher dose of methyl pentynol is required to produce the same effect equivalent to MS-222 (Howarnd and Schoettger, 1969). Sodium amytal is one of the most tried anaesthetics used for the transport of live fish. A concentration of 21 to 50 mg.l^{-1} of sodium amytal has been used to reduce the metabolic rate of the fry and live fish during transportation (Saha et al, 1955; Sreenivasan, 1962). However, sodium amytal has been used in soft water only as it is antagonistic to calcium. Benzocaine (Ethyl p-aminobenzoate) has ease of preparation, long storage life and safety (Laird and Oswald, 1975). Benzocaine has been used to anaesthetize salmonids, tilapia and catfish using doses between 33 to 200mg.l^{-1} (Ross and Ross, 1984).

The anesthetic effect of clove oil on the loach was the optimum when its dose was at 60-100 mg.l^{-1}, and it would not cause harm to the loach (Jie et al, 2018). Gangqiao et al, (2010) showed that when the dose of eugenol anesthetics was not higher than 27 mg.l^{-1}, the anesthetic effect on *Ictalurus punctatus* was the safest. When the dose of eugenol anesthetics was 5mg.l^{-1} and 8mg.l^{-1}, it would not produce an anesthetic effect on *Coilia ectenes* within 12 hours (Yinlong et al, 2016). 40mg.l^{-1} eugenol can produce an anesthetic effect on *Procambarus clarkii*.

The effective dose of MS-222 for *Acipenser schrenckii* was 80-100 mg.l^{-1} at 16~18°C and the fish could recover within 3-5 minutes (Jiayuan et al, 2011). The most suitable dose of MS-222 for bass was 20~50 mg.l^{-1} at 25°C. After 24 hours of anesthesia, the recovery rate could reach 100% in a clear water environment (Lijuan et al, 2015). When the dose of MS-222 was less than 40mg.l^{-1}, it had no anesthetic effect on the *Coilia macrognathos* Bleeker, but had a certain sedative effect (Yinlong et al, 2016). Xiaowang et al, (2017) showed that the safest dose of MS-222 was 60mg.l^{-1} for the transport of *Cynoglossus semilaevis*. Bin et al, (2018) found that MS-222 had no anesthetic effect on *Procambarus clarkii*. MS-222 had a good anesthetic effect on *Pelteobagrus fulvidraco* at 40-80 mg.l^{-1} (Lizhen et al, 2019). Yatao et al, (2019) also found that the dose of MS-222 anesthesia for bream was between 60 and 80 mg.l^{-1}.

Hypothermia is one of the oldest methods for anaesthestizing the fish but is not recommended generally. According to Mc Farland (1960), under anaesthetic treated condition, loss of reactivity to external stimuli and decrease in opercular frequency is regarded as the deep sedation stage of level of organization or I (2).

The present work suggests that the early developmental stages of *Channa punctatus* consumed less oxygen under anaesthetic conditions in comparison to controlled conditions. The weight-specific oxygen uptake of 3h Benzocaine exposed individuals was 423 % less than the controlled fishes while 24h treated *Channa* show 392 % reduction (Table-19). The opercular frequency also shows 43±3 % and 40±3 % reduction respectively in 3h and 24h exposure as compared to the controlled ones (Table-20). The results suggest that the reduction in oxygen uptake as well as opercular frequency by the *Channa punctatus* in 24h exposure is less pronounced than the 3h exposed fishes. This is probably due to some degree of acclimatization during 24h exposure. The variation both in oxygen uptake and opercular frequency due to 3h and 24h Benzocaine exposure was found to be highly significant. Similar findings have been reported by Jha and Ghosh (1993) in *Labeo rohita*. While working with the effect of Benzocaine in *Heteropneustes (=Saccobranchus) fossilis*, Pandit and Ghosh (1999) reported a significant reduction in their opereular frequency and aquatic oxygen uptake.

Goel and Sastri (1995) described that most of the anaesthetics decrease the sensitivity to respond to the chemicals, decrease in velocity of propagation and increase in the refractory period to impair the excitability of neurons. The aggregation of neurons within the medulla oblongata involved in respiratory function is known as the respiratory centre. The respiratory centre (located beneath the lower part of the floor of the fourth ventricle) is capable of coordinating breathing movements independently of the rest parts of the brain and spinal cord (Wang and Nagi, 1964), however, the boundary between the medulla oblongata and spinal cord is indistinct (Aronson, 1963).

Changlin et al, (2007) stated that the cerebral cortex is first inhibited by anesthetics, then act on the cerebellum and finally act on the spinal cord. Too much or too long exposure can make the anesthetics deep into the medulla, leading to shrinkage of respiration and blood vessel shrink and central paralysis, and ultimately lead to death.

Generally, the oxygen uptake of aquatic animals anesthetized by eugenol will rise first and then decrease. When the dose increased further, the oxygen uptake of *Acipenser baeri* decreased gradually, and obvious differences were occurring between different doses (Guangpeng et al, 2010). When the dose of eugenol was 10 $mg.l^{-1}$, the lowest oxygen uptake of *Trachinotus blochii* was 0.33 $mg.g^{-1}.h^{-1}$ (Chaohui et al, 2003) and when the optimum dose of eugenol for the carp anesthesia was 20 $mg.l^{-1}$, the oxygen uptake and ammonia excretion rate decreased significantly (Fulin, 2008).

When the individuals of *Channa punctatus* were introduced in Benzocaine mixed water, initially the fish starts secreting copious amounts of mucus, which forms a coat over the gills and reduces the efficacy of respiratory surfaces. In course of time, the fish gets acclimatized and the thick mucus coat is gradually lost, allowing better gaseous exchange. However, it remains low in comparison to the controlled one due to the reduced activity of the respiratory centre. *Channa punctatus* has small medulla oblongata (Munshi and Hughes, 1992). It seems that Benzocaine affects the medulla oblongata either directly or indirectly to show the reduced activity of the respiratory centre resulting in the reduction of oxygen uptake (Figure 3).

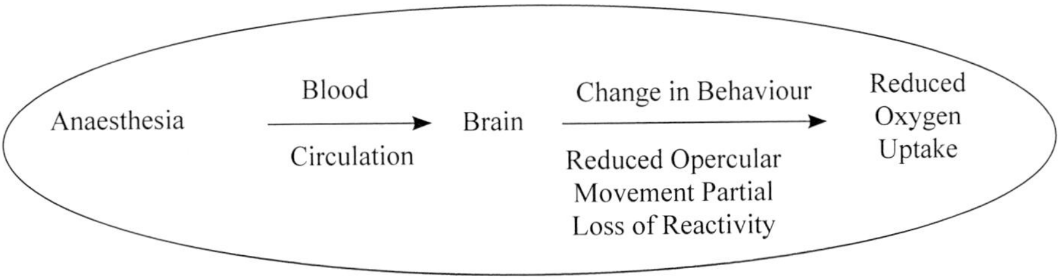

Figure 3: Model showing the probable course of action of the anaesthesia (Benzocaine) on the fish

The direct impact of the observed results revealed that anaesthetisation had marginal effects on survival rate, rather temperature and aeration had a more significant role. In addition, fish health also has a critical role in survivability during and after live transport. Hence the procedures prior to, during as well as after transportation play an important role in resolving the success of live-fish transports. Hence further studies need to be carried out in this regard for effective transportation protocols.

6

Effect of Ammonia on Gill and Oxygen Uptake

Introduction

Ammonia has been described as one of the most significant limiting factors for the growth and survival of aquatic living resources among water quality criteria (Russo and Thurston, 1991). Ammonia is a common pollutant of water bodies and forms the main nitrogenous waste in fishes (Tomasso et al, 1980). It enters the river and stream directly from livestock wastes and indirectly via poorly treated waste-water discharge. The chemistry of ammonia in freshwater has been extensively reviewed and its toxicity to the fish is generally expressed as ammonia concentration (EIFAC, 1970; WHO, 1986). Ammonia is produced mainly in the fish liver (Goldstein et al, 1982), it reaches the gills through the blood (Evans and More, 1988), excreted through the epithelium (Evans and Cameron, 1986) and finally released into the water body. The accumulation of nitrogen may induce alternations in branchial tissue and causes fish mortality (Peters et al, 1984; Woker and Wuhrmann, 1950).

Ammonia is a weak base, in the acidic range almost all the ammonia gets converted to ammonium ion (NH_4^+) but only a fraction of ammonia forms such ion at alkaline pH. However, in physiological ranges (pH: 7 to 8) about 2.5% of ammonia exists in a unionized form (Paley et al, 1993). The toxicity of ammonia in water is mainly the result of its unionized form (Richardson, 1997). The unionized ammonia accounts for the major part of metabolic ammonia excretion by diffusion down its concentration gradient (Cameron and Heisler, 1983). The unionized ammonia determines the toxicity level in the water and brings about great problems in pisciculture by causing various diseases (Burrows, 1964; Smart, 1976) and alternations in normal physiology (Cairns et al, 1975; Sarkar and Pramanik, 1986). The ratio of unionized ammonia to the total ammonia mainly depends on the pH and temperature of the freshwater. The concentration of unionized ammonia increases with increasing pH and temperature (Emerson et al, 1975). A rise in temperature and pH from 20^0C to 25°C and 7.5 to 8.5 respectively gives the fifteen-fold increase in the amount of unionized ammonia.

Numerous studies suggested that 4-5mg^{-1} of dissolved oxygen is the minimum amount that will support a large, diverse fish population. Ammonia is toxic to fish and aquatic organisms, even in very low concentrations. As levels near 2mg.l^{-1}, even ammonia tolerant fish like carp begin to die. Ammonia levels greater than approximately 0.1mg.l^{-1} usually indicates polluted water (Mohapatra and Saha, 2000). The danger that ammonia poses for fish depends on the temperature of water and pH, along with the dissolved oxygen and carbon dioxide levels. Higher the pH and warmer the temperature, the more toxic the ammonia is. Also, ammonia is much more toxic to fish and aquatic life when water contains very little dissolved oxygen and carbon dioxide. Carbon dioxide quickly combines in water to form carbonic acid which is a weak acid (Jhingran, 1975). The presence of carbonic acid may be good or bad depending on the pH and alkalinity of water. If the water is alkaline the carbonic acid will act to neutralize it. But if the water is already quite acid, the carbonic acid will only make things worse by making it even more acid.

The inhibitory effect of sublethal ammonia (120 g. l^{-1}) on the growth of fries of channel catfish (*Ictalurus punctatus*) was observed by Robinette (1976). The unfavourable influence of sublethal ammonia (110 g. l^{-1}) on the growth and metabolism of juvenile rainbow trout (*Salmo gairdneri*) was studied by Schulze-Wiehenbrauk (1976). Acute toxicity of ammonia and nitrite to the fries of cutthroat trout was investigated by Thurston and Russo (1978). The toxicity of ammonia on fingerlings of coho salmon was worked out by Buckley (1978), Anderson (1979) studied the effect of ammonia on the growth of juvenile Dover sole (*Solea solea*) and turbot (*Scophthalmus mexicum*). Exposure of ammonia to the developmental stages of *Salmo gairdneri* and its effect on growth was worked out by Calamari et al, (1981) and Soderberg et al, (1983) respectively. Effect of acute toxicity of ammonia to *Salmo gairdneri* and fathead minnows (*Pimephales promelas*) was investigated by Thurston and Russo (1983) and Thurston et al, (1983) respectively.

Das et al, (2004) studied the acute toxicity of ammonia and its sub-lethal effects on selected haematological and enzymatic parameters of a carp, *Cirrhinus mrigala* (Hamilton). An experiment was conducted by Bhakta (2006) to determine the 96h total ammonia tolerance level of four Indian freshwater fish species to prevent mortality caused by ammonia toxicity. Remen et al, (2008) studied the interactive effects of ammonia and oxygen on the growth and physiological status of juvenile *Gadus morhua*. Morphological changes in gills of juvenile Japanese flounder *Paralichthys olivaceus* (5.76±0.12 g) were investigated by Dong et al (2013) when fish were separately exposed to dissolved oxygen at 6.5 ± 0.5 mg.l^{-1} and supersaturated oxygen at 16.0 ± 2.0 mg.l^{-1} at different ammonia concentrations. Barbieri and Bondioli (2015) studied the acute toxicity of ammonia in Pacu fish, *Piaractus mesopotamicus* at different temperature levels. Susceptibility of Acute and chronic toxicity of ammonia in *Acipenser persicus*, fingerlings about the growth was studied by Raisei et al, (2019).

Brockway (1950) found that the oxygen content of the blood of trout decreases 1/7th of the normal value following ammonia exposure and suggested that ammonia may inhibit the ability of haemoglobin to combine with oxygen. Haemolysis in carp has been reported by Danecker (1964) due to exposure to ammonia. A decrease in circulating erythrocytes under ammonia exposure was also worked out in rainbow trout. While working on oxygen consumption of coho salmon, Sousa and Meade (1977) reported a reduction in the oxygen-carrying capacity of haemoglobin due to sublehtal exposure of ammonia in *Salmo gairdneri* while its acute lethal concentration increases oxygen uptake and heartbeat. Recently, Linton-Tyler et al, (1998) have studied the effect of sublethal concentration ammonia on certain physiological aspects viz., metabolic cost and winter warming etc.

The observations of Lloyd (1961) suggested that ammonia may act system primarily on the respiratory system. The toxic and subtoxic anatomical and histo-pathological effect of ammonia water on gills in *Cyprinus carpio* was studied by Flis (1967, 1968a and 1968b). In another experiment, Smart (1976) found minor histopathological changes at an acute lethal concentration which impaired oxygen uptake.

Among the Indian air-breathing fishes, the effect of ammonium sulphate has been studied on oocyte, testis, liver and thyroid respectively of *Channa punctatus* by Ram and Sathyanesan (1986, 1987a and 1987b). Banerjee and Paul (1993) studied the effect of ammonia on the skin while Paul and Banerjee (1995) on the opercular epithelium of *Heteropneustes (= Saccobranchus) fossilis*. Joy and Sathyanesan (1977) investigated the effect of ammonia on the thyroid of *Clarias batrachus*. Saha and Ratha (1994) studied the induction of the ornithine-urea cycle in a freshwater teleost, *Heteropneustes fossilis*, exposed to high concentrations of ammonium chloride.

As the effect of ammonia on the respiratory organs of air-breathing fishes is probably limited, therefore, in the present work, an attempt has been made to elucidate the effect of ammonia on the structure and function of the gills of *Channa punctatus*.

Materials and Methods

Proper sanitization of all the equipment and accessories is one of the essential steps for the study of gill structure and determination of oxygen uptake of fishes under ammonia toxication in this pandemic of COVID-19.

Live *Channa punctatus* (Bodyweight=6.0 to 10.0 g and Total length=8.5 to 10.6 cm) of either sex were kept in a large plastic pool (capacity: 1000 litre) in the Aquarium House of the Department of Zoology, T. M. Bhagalpur University, Bhagalpur, and were treated with methylene blue for about one week. Then, they were transferred to the Laboratory for acclimatization in a plexiglass aquarium (capacity: 40 litre) for a week. During acclimatization, the water was removed on every alternate day and the fishes were fed with trash fish and chopped goat liver. The temperature, pH and dissolved oxygen of water were 27.0 ± 1.0°C, 7.3±0.2 and 7.40±0.40 $mg.l^{-1}$ respectively.

SQ grade (99% pure) ammonium chloride (NH_4CI) of Glaxo Laboratories (India) Limited, Bombay was used in this work to get ammonia in water. A test solution of ammonia of desired concentration was prepared from its 15% (8.00g NH_4Cl. l^{-1}) stock solution. To determine 96h-LC_{50} dose of ammonia, individually weighed ten isomorphic fishes were kept separately in smaller plexiglass aquaria (30×30×30 cm) in different doses of freshly prepared ammonia solutions. Reactions of fishes to ammonia were observed during the first one hour followed by every half an hour up to eight hours and then at an interval of 24h, 48h and 96h. Ammonium acetate was added as a buffer to check the variation of pH (Svehla, 1985). Cessation of opercular movement and lack of response to prodding with a glass rod was considered as criteria of death. The per cent mortality and probit mortality were plotted on graph paper against log concentrations. The 96h-LC_{50} 0of ammonia was calculated by both graphical interpolation and probit analysis method (Finney, 1964).

To study the effect of sublethal dose (one-third of 96h-LC_{50}) of ammonia on gills of *Channa punctatus*, the fish body weights 6.20 to 8.30g were divided into three groups (A, B and C) with five fishes in each group. Group C was kept in tap water as control or normal while A and B were kept separately in experimental water with ammonia at pH: 7.7 and 6.5 respectively. 0.25ml of glacial acetic acid was added to the solution of ammonia to lower the pH from 7.7 to 6.5. The pH of the experimental solution of ammonia and controlled water was monitored by 'C-TRONICS' portable pH meter. The solution was replenished daily. After a stipulated period of exposure, the fishes were stunned and gills were dissected out both from control and experimental fishes.

For Scanning Electron Microscopic work, the gills were perfused with 2.5% glutaraldehyde in 0.1 M phosphate buffer at pH: 7.4. The dissected gills were kept overnight in the fixative at 4°C. The tissue was then washed in 0.1 M phosphate buffer (pH: 7.4), dehydrated in graded ethyl alcohol, a mixture of ethyl alcohol and acetone and finally kept in anhydrous acetone at room temperature. Dehydrated specimens were subjected to critical point drying in CPD equipment (Polaron-Watford-England) using then fixed on liquid carbon dioxide as transitional gas. The dried tissues were then fixed on aluminum stubs with double adhesive tape and conducting fluid (Silver dag) and finally sputtered with gold (160-165Å) in a sputter coater (Edwards, IS-150, England). The specimens were examined in Scanning Electron Microscope, PSEM-500 (Philips, Holland) at Regional Sophisticated Instrumentation Centre (RSIC), Bose Institute, Calcutta and in JEOL–JSM 5200 at University Scientific and Instrumentation Centre (USIC), Jadavpur University, Jadavpur and photographs were taken for further studies.

To study aquatic oxygen uptake of *Channa punctatus*, a cylindrical glass respirometer was used (Ghosh and Munshi, 1987). Details of the experimental set up have already been given in chapter 3 of this book. To measure the oxygen uptake of control fishes, the reservoir was first filled with tap water. The fish was then introduced into the respirometer and then it was also filled with the same water. The fish was allowed to acclimatize in the respirometer for an hour and then the readings were taken at an interval of 30 minutes each. Winkler's volumetric method (Welch, 1948) was used to measure the dissolved oxygen level of inspired and expired water. To study the effect of ammonia on oxygen uptake and opercular frequency, ammonia exposed (24h, 48h and 96h) fishes were introduced in the respirometer which was filled with the experimental solution of ammonia. The fish was acclimatized in the respirometer and then the oxygen uptake was determined. The opercular movement was counted for 5 minutes both in controlled as well as treated fishes during experimentation and an average value was calculated to determine the mean opercular frequency (min^{-1}).

The data were analyzed with the help of Graphpad prism 5.04software. The data of 96h–LC_{50} determination, oxygen uptake and opercular frequency (min^{-1}) have been tabulated in Tables–21 and 22.

Observations and Results

Determination of 96h LC_{50}: *Channa punctatus* (body weight=6.20 to 8.30g) were exposed to different concentrations of ammonia to determine the 96h LC_{50} value. While no mortality was observed up to 1372 mg-NH_3.l^{-1}, a mortality of 30 and 100 % was observed in 1430 and 1830 mg-NH_3.l^{-1} respectively (Table-21). When the per cent mortality of fishes was plotted against log concentration of ammonia, a sigmoid curve was obtained. The graphical plot suggested LC_{50} value for ammonia to be 1428 mg.l^{-1} for the fishes of body weight 6.20 to 8.30g. When per cent mortalities were changed to probit kill and data were plotted against log concentration, a straight line was obtained. The respective LC_{50} value for *Channa punctatus* was found to be 1462 mg.l^{-1}.

Observations on Behaviour: The behaviour of *Channa punctatus* to ammonia was more or less similar in both the experimental solutions (pH: 7.7 and 6.5). As soon as the fishes were introduced in the experimental solutions, they became excited and showed erratic swimming for about 20-30 minutes. The fishes sometimes tried to jump outside the aquarium and then gradually settled down at the bottom. The opercualr movement increased which gradually lowered, however it remained more than the controlled one. While the fishes in controlled water mostly remained towards the surface, the fishes in the experimental solution of ammonia preferred to stay at the bottom of the aquarium. An enormous amount of mucus was seen over the body.

Observations of the Gills: *Channa punctatus* has four pairs of functional gills for aquatic respiration in addition to accessory respiratory organs. There is a gradual reduction in the relative size from the first to the fourth gill. Each gill arch bears rows of stiff gill rakers on its pharyngeal side. The gill filaments project from both sides of the gill arch as finger-like structures having alternately arranged thin and leaf-like lamellae on both of its sides.

SEM of Normal (Control) Gills: Under Scanning Electron Microscope, the gill arch was found to be lined with epithelium having either pentagonal or hexagonal epithelial cells with sporadic mucous gland openings. The epithelial cells are covered over by the folding of plasmalemma or micro ridges; hence the name micro ridged epithelial cells (Plate-III, Figure-3). The sparsely situated gill rakers are short and stumpy with numerous denticles over it (Plate-III, Figure-1). The epithelium of the gill raker was found to be similar to the epithelium of the gill arch. The gill filament was covered with epithelium characterized by the presence of compact micro ridged epithelial cells having comparatively fewer mucous gland openings (Plate-II, Figure-4). Each filament bears lamellae up to their tip region. The tips of filaments are free and not fused (Plate-III, Figure-2). Each lamella is projected at a right angle to the long axis of filament and situated parallel to the adjacent lamellae. The epithelial surface of the lamellae appeared uneven in its general view (Plate-III, Figure-5). The surface of secondary lamellae revealed the presence of pentagonal epithelial cells with either feeble or broken micro ridges or a few minute mucous gland openings (Plate-III, Figure-6).

SEM of Ammonia Treated Gills (at 24 hours): Both in alkaline and acidic solutions of ammonia, the gill arch and gill raker showed wrinkling of the epithelium (Plates-IV and V. Figure-1). The gill arch epithelium was found to be more folded towards the base of gill filaments. The folding was more conspicuous in the alkaline

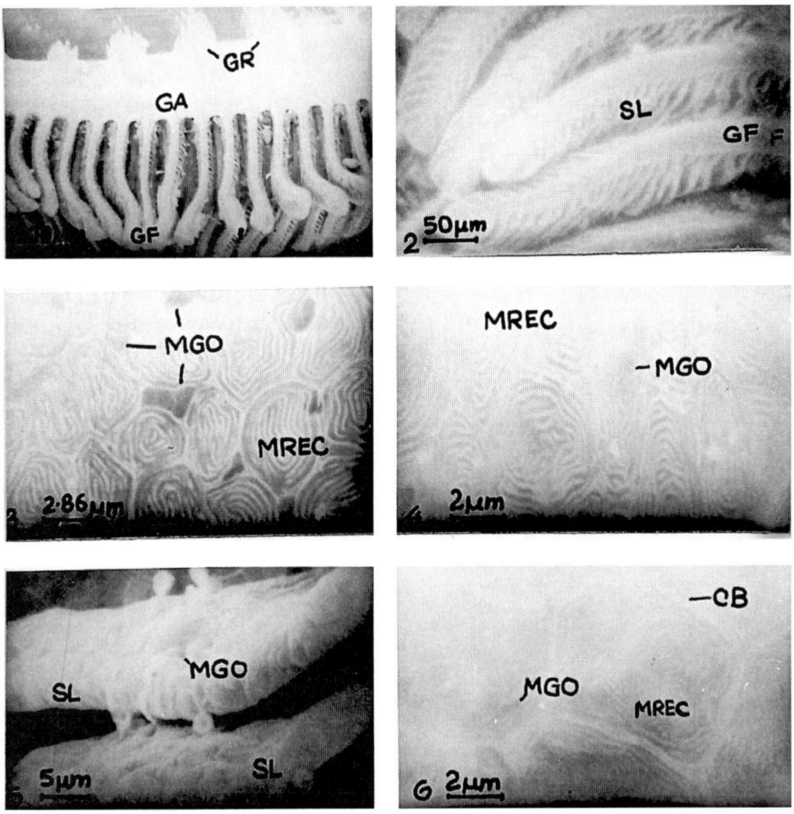

Explanation of Plate – III

SEM Photomicrographs of normal gill of *Channa punctatus*

1. Part of first gill arch showing gill rakers and gill filaments X50.
2. Part of gill arch showing tips of few filaments X200.
3. Gill arch surface showing micro ridged epithelial cells and mucous gland openings X3500.
4. Gill filament surface showing micro ridged epithelial cells and few mucous gland openings X5000.
5. Two adjacent lamellae showing corrugated surface and minute mucous gland openings X2000.

6. Secondary lamellar showing less prominent micro ridged epithelial cells and few mucous gland openings X5000.

Abbreviations Used

CB = Cell boundary, GA = Gill arch, GF = Gill filament, GR = Gill raker, MGO = Mucous gland openings, MREC = Micro ridged epithelial cells and SL = Secondary lamella..

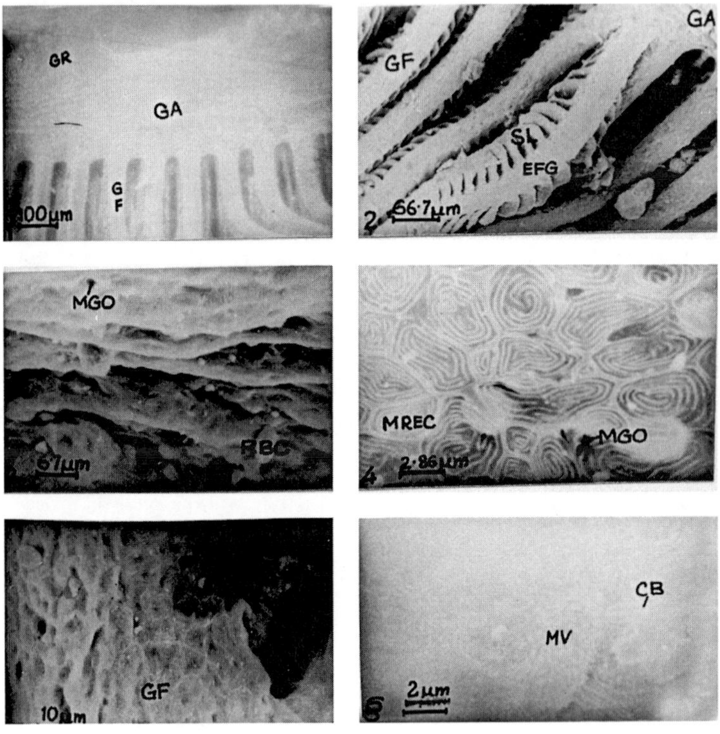

Explanation of Plate – IV

SEM Photomicrographs of the gills of *Channa punctatus* exposed to alkaline ammonia solution for 24 hours.

1. Part of gill arch showing two gill rakers and few filaments with wrinkled epithelium X100.
2. Part of gill arch showing extra filamental growth and wrinkling of epithelium X150.
3. Lower part of gill arch just above the gill filaments showing wrinkling, fusion and occasional rupture of epithelium X1500.
4. Part of gill arch surface showing well defined micro ridged epithelial cells and mucous gland openings X3500.

5. Part of gill filament and secondary lamellae showing micro ridged epithelial cells and mucous gland openings X1000.
6. Secondary lamellar surface showing epithelial cells with irregular margins and micro ridges broken into micro villi X5000.

Abbreviations Used

EGF = Extra Gill filament, GA = Gill arch, GF = Gill filament, GR = Gill raker, MGO = Mucous gland openings, MREC = Micro ridged epithelial cells, MV = Micro villi and SL = Secondary lamella.

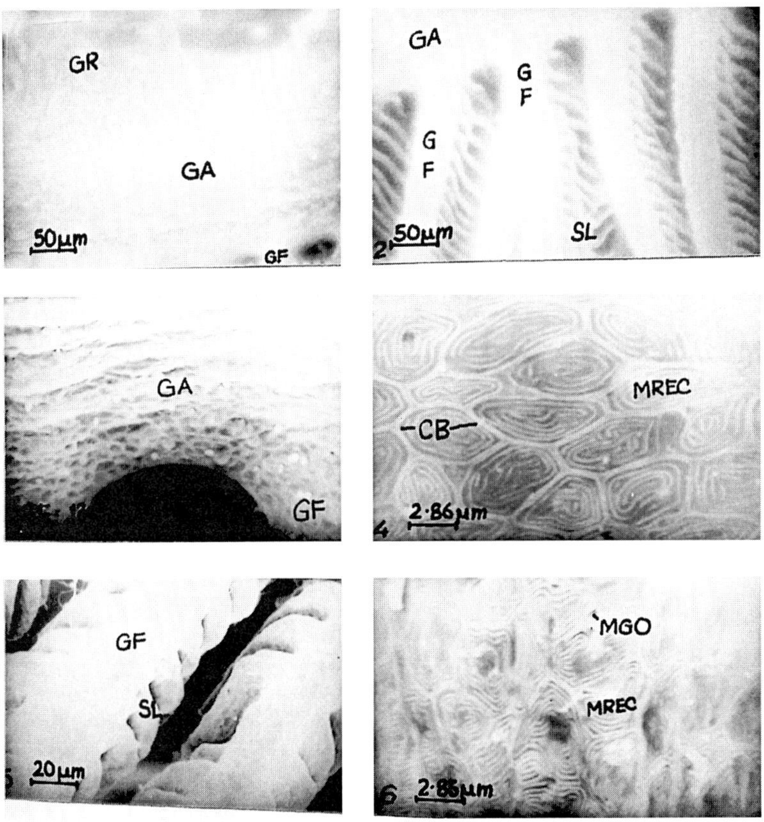

Explanation of Plate – V

SEM Photomicrographs of the gills of *Channa punctatus* exposed to acidic ammonia solution for 24 hours.

1. Part of gill arch showing two adjacent gill rakers with wrinkled epithelium and base of few filaments X200.
2. Few filaments of which one showing bifurcation of filament. The epithelium showing wrinkling with swelling X200.

3. Base of gill arch showing shrinkage, folding and shrunken epithelium X750.
4. Gill arch showing micro ridged epithelial cells and mucous gland openings X3500.
5. Part of two adjacent gill filaments showing loss of secondary lamellae X500.
6. Gill filament surface showing shrunken micro ridged epithelial cells and mucous gland opening X3500.

Abbreviations Used

BOF = Bifurcation of filament, GA = Gill arch, GF = Gill filament, GR = Gill raker, MGO = Mucous gland openings, MREC = Micro ridged epithelial cells and SL = Secondary lamella.

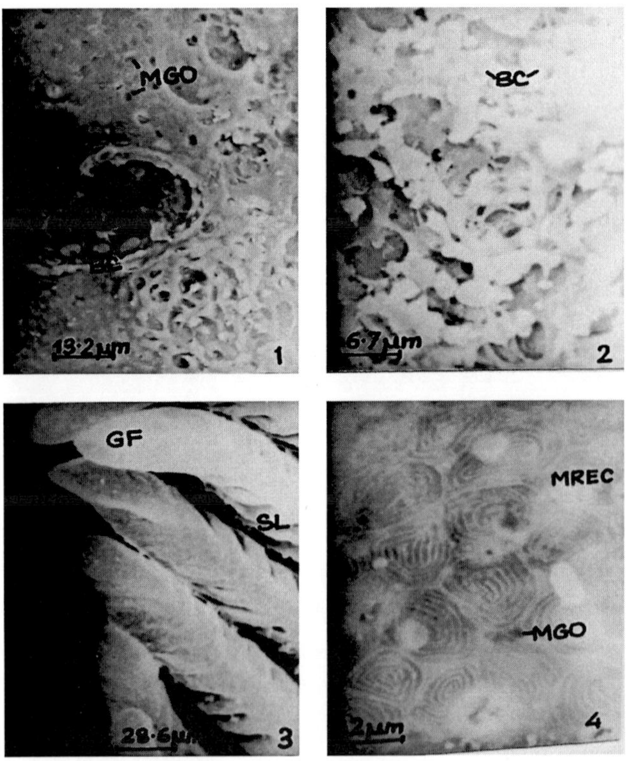

Explanation of Plate – VI

SEM Photomicrographs of the gills of *Channa punctatus* exposed to alkaline ammonia solution for 48 hours.

1. Part of gill arch surface showing damaged epithelial surface with exposed blood cells and numerous mucous gland openings X 750.
2. Part of gill arch showing complete removal of epithelium and exposed blood cells X 1500.
3. Few adjacent filaments showing fused secondary lamellae at their tips X 350.
4. Gill filament surface showing micro ridged epithelial cells and mucous gland opening X5000.

Abbreviations Used

BC = Blood cells, GA = Gill arch, GF = Gill filament, MGO = Mucous gland openings, MREC = Micro ridged epithelial cells and SL = Secondary lamella.

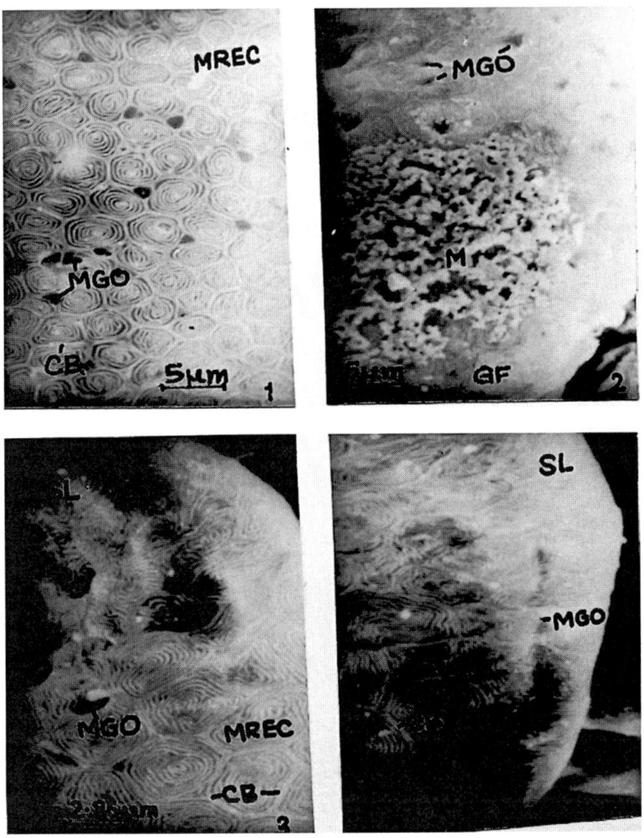

Explanation of Plate – VII

SEM Photomicrographs of the gills of *Channa punctatus* exposed to acidic ammonia solution for 48 hours.

1. Part of gill raker surface showing mucous gland openings and micro ridged epithelial cells at their base X 2000.
2. Gill filament surface showing some abnormality and release of mucous X2000.
3. Part of gill filament with secondary lamellae showing well defined micro ridged epithelial cells its surface X3500.
4. Single secondary lamella showing micro ridged epithelial cells over its surface X3500.

ABBREVIATIONS USED

GF = Gill filament, GR = Gill raker, MGO = Mucous gland openings, MREC = Micro ridged epithelial cells,

SL = Secondary lamella.

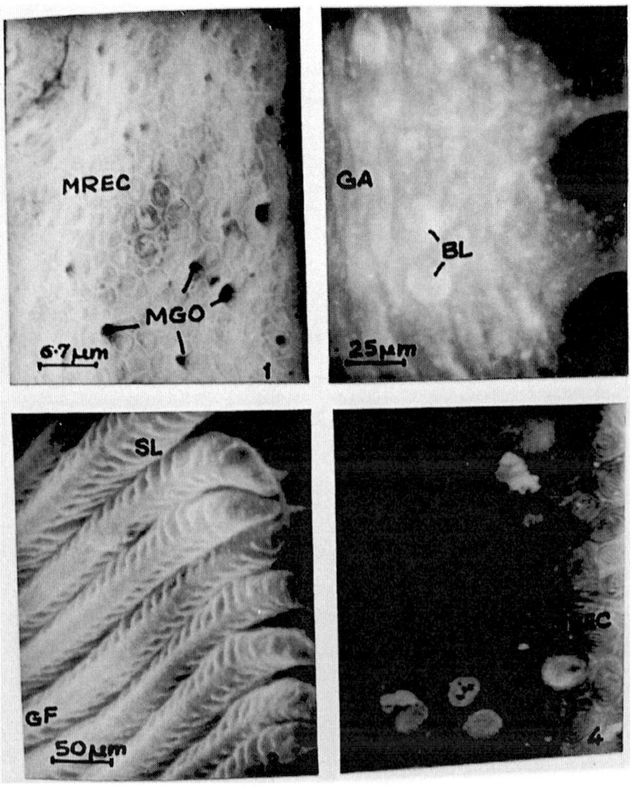

Explanation of Plate – VIII

SEM Photomicrographs of the gills of *Channa punctatus* exposed to alkaline ammonia solution for 20 days.

1. Gill arch surface with micro ridged epithelial cells and numerous mucous gland openings X1500.
2. Part of gill arch surface showing shrinkage, swelling or nodule etc. X 400.
3. Few adjacent filaments showing bleb or nodules and their tips with fused secondary lamallae X200.
4. Gill arch surface showing rupture, some elevated cells, micro ridged epithelial cells and mucous gland opening X 2000.

ABBREVIATIONS USED

BL = Bleb or nodule, GA = Gill arch, GF = Gill filament, MGO = Mucous gland openings and MREC = Micro ridged epithelial cells.

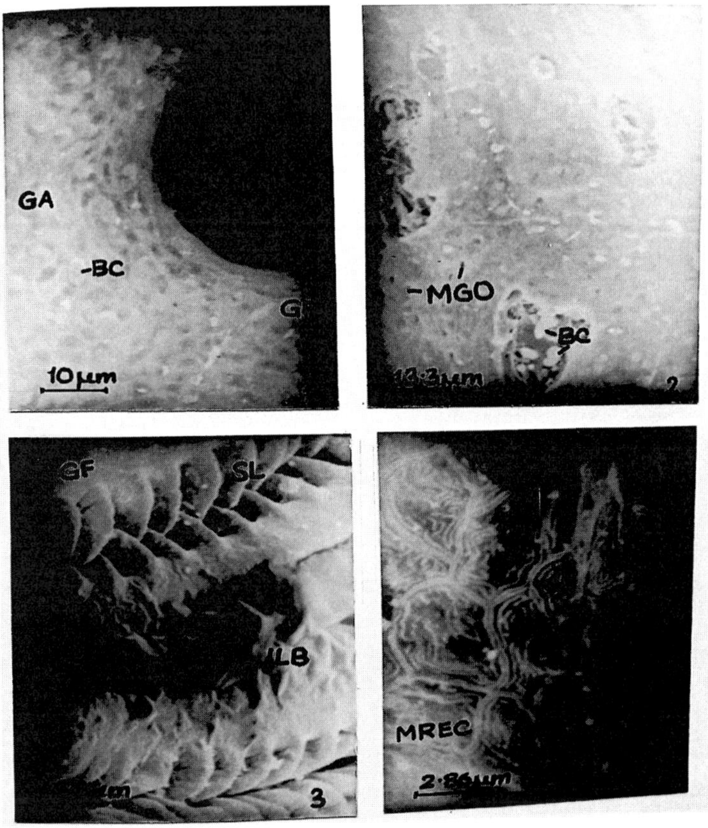

Explanation of Plate – IX

SEM photomicrographs of the gills of *Channa punctatus* exposed to acidic ammonia solution for 20 days.

1. Part of gill arch and gill filament surface showing damaged epithelium and leakage of blood cells X 1000.
2. Gill arch surface showing rupture and loss of epithelium, appearance of blood cells and numerous mucous gland openings X 750.
3. Two adjacent filaments fuse of form inter lamellar bridge X 500.
4. Filamental surface from tip region showing loss of micro ridges of broken micro ridges in the epithelial cells and few mucous gland openings near the marginal channel of the lamellae X 3500.

ABBREVIATIONS USED

BC = Blood cells, GA = Gill arch, GF = Gill filament, ILB = Inter lamellar bridge, MGO = Mucous gland openings and MREC = Micro ridged epithelial cells. Solution of ammonia (Plate- IV, Figure-3) as compared to the acidic solution of ammonia (Plate-V, Figure-3).

Table 21: Mortality of *Channa punctatus* at different concentrations of ammonium chloride after 96h of exposure.

Concentration of (g)	Log concentration	Number of fishes exposed	Number of fishes dead	Percentage mortality	Probit kill
4.30	0.635	10	--	--	--
4.50	0.653	10	3	30	4.48
4.75	0.677	10	7	70	5.52
5.00	0.699	10	8	80	5.84
5.75	0.760	10	10	100	8.72

The micro ridged epithelial cells and the mucous gland openings of fishes were also prominent in the epithelium of alkaline ammonia-treated gills (Plate-IV, Figure-4) in comparison to the epithelium of acidic ammonia solution (Plate-V, Figure-4). Besides, the lower part of the gill arch between two adjacent filaments showed shrunken epithelium in acidic pH (Plate-V, Figure-3). Few filaments showed bifurcation within it (Plate-V, Figure-2). At alkaline ammonia solution, an additional outgrowth of filaments was observed which fills up the inter-filamental space (Plate-IV, Figure-2). While the filaments show lamellae up to their tips in alkaline ammonia-treated gills, the lamellae of the tip region showed fusion in acidic ammonia solution. The gill filament surface showed numerous mucous gland openings and micro ridged epithelial cells with the frizzy appearance of the micro ridges in both the conditions (Plates-IV and V, Figure-5). However, the above changes are more prominent in the alkaline solution of ammonia. Partial loss and damage of few lamellae were observed in the acidic solution of ammonia (Plate-V, Figure-5). The lamellar surface showed loss of definite cell boundaries and breaking of micro ridges into villi like structures in alkaline ammonia-treated gills (Plate-IV, Figure-6). Contrary to this the micro ridged epithelial cells of acidic ammonia showed invagination in few cells to cause depressions therein (Plate- V, Figure-6).

SEM of Ammonia Treated Gills (at 48 hours): The gill arch surface showed damage and loss of epithelial cells in the alkaline solution of ammonia resulting in the exposure of some erythrocytes and rupture of few mucous gland openings (Plate-VI, Figures-1 and 2). Some of the micro ridged epithelial cells lost their micro ridges. The filament surface also showed similar changes. On the other hand, in acidic ammonia solution, the gill arch surface showed the presence of numerous mucous gland openings and micro ridged epithelial cells with beautifully arranged whorls of micro ridges (Plate-VII, Figure-1). Similarly, the filament surface also showed well defined micro ridged epithelial cells (Plate-VII, Figure-3). Both in the gill arch and filament surface, numerous mucous gland openings were observed interspersed between micro ridged epithelial cells (Plate-VII, Figure-2). However, the number of mucous gland openings was more on the filament surface in comparison to the gill arch surface. At places, a deposit of mucus was seen on the filament surface (Plate-VII, Figure-2). Complete fusion of lamellae at the tips of filaments was seen only in the alkaline ammonia solution (Plate-VI, Figure-3). The lamellae did not show much change in their surface ultrastructure in both the solutions (alkaline and acidic) of ammonia (Plates- VI and VII, Figure-4).

SEM of Ammonia Treated Gills (at 20 days): In an alkaline solution of ammonia, the gill arch surface showed well defined micro ridged epithelial cells, and numerous mucous gland openings (Plate-VIII, Figure-1). However, in acidic ammonia solution, the epithelium had ruptures in it (Plate-IX, Figure-2) and the micro ridged epithelial cells were seen depressed giving cavity like appearance. Few mucous gland openings were observed in both the solutions of ammonia (Plates-VIII and IX, Figure-1). The filament surface was found to be more or less similar to the gill arch surface. In a few filaments, some nodule or bleb like elevations or evaginations (Plate-VIII, Figure-2) was observed in the alkaline solution of ammonia, but there was no lamellar abnormality. Under acidic ammonia solution, six to seven lamellae of two adjacent filaments fused to form interlamellar bridges (Plate-IX, Figure-3). The lamellar surface was again more or less similar in both the solutions. Occasionally some minute mucous gland openings were also observed (Plates-VIII and IX, Figure-4).

Aquatic oxygen uptake: The mean value of oxygen uptake per unit time for a controlled (normal) *Channa punctatus* of average body weight 7.176±0.491g was recorded to be 0.1379±0.0340 $mlO_2.h^{-1}$. The oxygen uptake increased to 0.3024±0.0719 and 0.52191±0.0729 $mlO_2.h^{-1}$ for fishes of similar body weight at a sublehtal dose of ammonia for 24h and 48h respectively. In the

96h ammonia-treated fish of the same body weight, the oxygen uptake was recorded to be only 0.2653±0.0710 $mlO_2.h^{-1}$ (Table-22). Similarly, the weight-specific oxygen uptake of a control fish was calculated to be 0.0192±0.0040 $mlO_2 g^{-1}.h^{-1}$, whereas it was 0.0435±0.0118, 0.0703±0.0090 and 0.0285±0.0086 $mlO_2 g^{-1}.h^{-1}$ in 24h, 48h and 96h exposed fishes respectively. In comparison to the control fish, the oxygen uptake per unit time in 24h and 48h ammonia exposed fish increased by 119.23 and 278.40% respectively. However, the increase was only 48.86% at 96h exposed fish.

In a control fish, the opercular frequency was recorded to be 37.0±3.0 min^{-1} which increased to 44.0±3.0 and 50.0±4.0 in 24h and 48h exposure of ammonia respectively. However, at 96h exposure of ammonia, the opercular frequency came to 43.0±4.0 min^{-1} (Table-22). An increase of 18.92%, 35.14% and 16.22% was calculated in the opercular frequency during 24h, 48h and 96h of treatment respectively when compared to the opercular frequency of a control fish.

The oxygen uptake per unit time, per unit body weight and the opercular frequency at 48h exposure was found to be 72.60%, 61.46% and 13.64% more in comparison to the corresponding parameters at 24h exposure of *Channa punctatus*. Further, at 96h exposure, oxygen uptake per unit time, per unit body weight and the opercular frequency showed 32.10%, 34.65% and 2.27% reduction in comparison to 24h exposure. However, these parameters showed 56.83%, 59.52%and 14.00% reduction when compared with 48h exposed fish.

The changes in oxygen uptake and opercular frequency due to exposure of ammonia were highly significant at 48h ($p<0.001$) followed by 24h ($p<0.01$) and 96h ($p<0.05$) of ammonia exposure.

Table 22: Aquatic oxygen uptake ($mlO_2.h^{-1}$ and $mlO_2.g^{-1}.h^{-1}$) and opercular frequency (min^{-1}) in controlled and ammonia exposed *Channa punctatus* at 28±1° C.

Average weight of a fish (g)	Control			24h Exposure			48h Exposure			96h Exposure		
	Opercular frequency (min^{-1})	Oxygen uptake		Opercular frequency (min^{-1})	Oxygen uptake		Opercular frequency (min^{-1})	Oxygen uptake		Opercular frequency (min^{-1})	Oxygen uptake	
		$mlO_2.h^{-1}$	$mlO_2.g^{-1}.h^{-1}$		$mlO_2.h^{-1}$	$mlO_2.g^{-1}.h^{-1}$		$mlO_2.h^{-1}$	$mlO_2.g^{-1}.h^{-1}$		$mlO_2.h^{-1}$	$mlO_2.g^{-1}.h^{-1}$
				**	**	**	***	***	***	*	*	*
7.176 ± 0.491	37.0 ± 3.0	0.3179 ± 0.0340	0.0192 ± 0.0040	44.0 ± 3.0	0.3024 ± 0.0719	0.0435 ± 0.0118	50.0±4.0	0.5219 ± 0.0729	0.0703 ± 0.0090	43.0±4.0	0.2053 ± 0.0710	0.0285 ± 0.0086
				(18.92)	(119.23)	(126.79)	(35.14)	(278.40)	(276.40)	(16.22)	(48.86)	(48.22)
							{13.64}	{72.60}	{61.46}	{-2.27}	{-32.10}	{-34.65}
										[-14.00]	[-56.83]	[59.52]

***** $p<0.001$, ** $p<0.01$, *$p<0.05$; () = % increase in comparison to control; { } = % increase/ decrease in comparison to 24 h exposure; [ ] = % decrease in comparison to 48 h exposure**

Discussion

Fish gills are complex organ, which performs various functions including gaseous exchange, ionic regulation (Eddy, 1982), osmoregulation, acid-base regulation (Randall et al, 1982) and regulation of circulating hormones (Nekvasil and Olson, 1986; Olson et al, 1986) etc. In nature, gills are subjected to the water of different physicochemical character and fishes allow their branchial tissue to cope up with exposed variations in the medium. Changes in the structure and function of gills under the exposure of different types of substances such as heavy metals (Kumar and Pant, 1981; Singh and Munshi, 1996), Pesticides (Munshi and Singh, 1971; Roy and Munshi, 1988; Dutta et al, 1992), an alkaloid (Roy et al, 1986) and seawater (Jha, 1995; Adhikari et al, 1997) etc. have been observed. The toxic effect of ammonia has been studied by Tomasso et al, (1980); Peters et al, (1984), Ostrensky and Brugger (1992) etc. Some workers have reported that lethal and sublethal ammonia causes severe changes in gill structure (Kuhn and Koecke, 1956; Burrows, 1964; Reichenback-Klinke, 1967; Flis, 1967 and Smart, 1976), decrease in oxygen content of the blood (Brockway, 1950), decrease in circulating erythrocytes (Reichenback-Klinke, 1967) haemolysis (Danecker, 1964) and variation in carbohydrate metabolism (Reddy et al, 1996) etc. To observe a histopathological effect, chronic exposure to ammonia was worked out by Wajsbrot et al, (1993) in the juveniles of *Sparus aurata* (gillhead seabream).

Toxicants are known to alter the oxygen uptake of fish. Jones (1964) and Ranjeet et al, (2013) suggested that fish responds to toxic materials by increased or decreased opercular movements. Strzyzewska et al, (2016) studied morphologic evaluation of the gills in fish and pollution in the aquatic environment. Miramontes et al, (2020) investigated the effects of ammonia toxicity in fish, and other animal species.

LC50 or TLm (Median Tolerance limit) is the least concentration in which 50% of the organisms died, which is tested for a specific period of exposure. It is useful in the calculation of maximum allowable toxicant concentration (MATC). The LC50 value for ammonia in fishes from different habitats ranged between 0.16mg.l^{-1} to 2000.00 mg.l^{-1}. The differences are probably due to variations in experimental temperature and pH, bodyweight of fishes, nature of water, fish species etc. Thurston et al, (1983) found a range of 0.75 to 3.40 mg.l^{-1} of ammonia to be the 96h LC50 in fathead minnow (*Pimephales promelas*). It was calculated to be 0.16 to 1.1 mg.l^{-1} in *Salmo garidneri* by Thurston and Russo (1983). Cardoso et al, (1996) found 48h LC50 as 0.48±0.0 and 0.92±0.07 mg.l^{-1} for larvae (0.02g) and alevians (0.411 ± 0.11g) of *Lophiosilurus alexandri*. Richardson (1997) reported that 96h LC50 at 15°C and pH 7.5 to 8.1 ranged between 0.75 to 2.35 mg.l^{-1} for *Anguilla dieffenbanchii* and *Anguilla australis*. Saha et al, (1956b) reported that the spawn of Indian major carps can tolerate 2.5 mg.l^{-1} of free ammonia and 15 ppm of dissolved ammonia in the form of inorganic salts. Reddy et al, (1996) determined a 48h LC50 value of ammonia to be 2.5 ppm for *Cyprinus carpio* of body weight 250±10 mg. However, a 96h LC50 value of ammonia was calculated to be 2000.0 mg.l^{-1} by Paul and Banerjee (1995, 1996) for *Heteropneustes fossilis* of bodyweight 45-50g. In the present work, 96h LC50 value for ammonia on *Channa punctatus* (body weight 6.20 to 8.30 g) was found to be 1478 mg.l^{-1} (4.65g-NH_4Cl.l^{-1}). The LC50 value of ammonia in *Channa punctatus* was higher than the dose of other reported fishes except *Heteropneustes fossilis*. The higher dose in these fishes is probably due to their dual breathing nature. These air-breathing fishes are quite hardy as compared to other aquatic breathing fishes owing to their capability to sustain in the adverse ecological habitat.

Results obtained from the study of Bhakta (2006) demonstrated a difference in ammonia tolerance of four freshwater fish species. 96hr-LC_{50} were 15, 10, 25 and 30 mg/l in *Catla catla, Labeo bata, Cyprinus carpio* and *Tilapia mossambica* respectively which indicated that ammonia tolerance level was maximum in tilapia which followed by common carp, catla and bata, respectively.

Effect of ammonia on gill structure: The fish gill is easily damaged even by the low concentration of pollutants. It is, therefore essential to have an idea of the normal morphology of gills to permit an assessment of alterations. The Scanning Electron Microscope enhances the visualization of surface morphology. The surface ultrastructure of normal gills of *Channa punctatus* is similar to that of the description of Singh (1990), Munshi and Singh (1992) and Singh and Munshi (1996). The surface ultrastructure showed differences in the topographical details of different gill components of *Channa punctatus*. While gill arch and gill raker surfaces are provided with polygonal (pentagonal or hexagonal) epithelial cells with numerous mucous gland openings and taste bud openings, the gill filament possesses compact micro ridged epithelial cells with fewer mucous gland openings but no taste bud openings. The lamellae showed pentagonal epithelial cells with broken micro ridges and few mucous gland openings. Such a variation and gradual reduction of micro ridges from gill head to gill filament to lamellae have also been shown by Lewis (1979) in *Clarias batrachus* and by Ghosh (1984) in *Anabas testudineus* respectively. The distinction between micro ridges of gill filament and microvilli of lamellae is well defined. The microvilli appeared as if they are broken parts of the micro ridges. Olson and Fromm (1973) and Rajbanshi (1977) are also of the opinion that microvilli are ridges of varying lengths and patterns of foldings of micro ridges. The microvilli of lamellae probably protect it from polluted water of habitat where the fishes thrive.

Various ideas have been given regarding the functional significance of micro ridges and therefore, it is difficult to pinpoint a single function. According to Hughes and Wright (1970) and Hughes (1979), micro ridged epithelial cells help in holding the mucus over the gill surface; it will increase the surface area to facilitate gaseous exchange in water and the formation of micro-turbulence in ventilating water to minimize its velocity. The absorptive and secretory activities of micro ridges have been proposed by Jones, et al, (1996) and Olson and Fromm (1973). It may provide a reserve surface area for stretch (Wassersug and Johnson, 1976) and spread of mucus away from goblet cells (Sperry and Wassersug, 1976). Microridges are directly associated with mucous secreting cells and the release of mucus. Thus, the view suggesting anchorage and spreading of mucus seems reasonable. It seems that the micro ridges are selective in their function at different parts of the gills.

Under normal conditions, the gills are covered over with a thin layer of mucus. The mucus has many fold functions viz. it helps in keeping filaments clear from mud and fine sediments, protects plasmalemma from damage by abrasion, retards infections through its antiseptic properties and prevents attachment of parasites as it acts as a lubricant. The mucous glands may change their function according to the need of the fish. Pisam et al, (1990) explained that mucous cells excrete ions in marine water but accumulate it in freshwater conditions. As *Channa punctatus* usually live in the derelict water of swamps and ditches, functions like keeping filaments clear, acting as antiseptic and prevention of attachment of parasites etc. seem more logical.

Usually, the gill rakers are modified according to the food and feeding habit of the fish. Short, stumpy and sparsely distributed gill rakers of *Channa punctatus* suggest the carnivorous nature of the fish (Khanna, 1998). The observation corroborates with the findings of Munshi and Hughes (1992) in which they discussed that *Channa punctatus* is a column feeder with a high predatory habit. Munshi et al, (1984) observed short and stumpy gill rakers in carnivorous fishes, which are suited for larger food particles. Norman (1975) also discussed that short and sparsely situated gill rakers are found in those fishes, which engulf larger prey. The gill rakers might also have protective functions against mechanical hazards (de Graff et al, 1987). The presence of taste buds at the base of the gill raker helps the fish in the gustatory sense so that the fish might select or reject a particular food item. It also acts as an effective chemical indicator. It enables the fish to detect the presence of toxins in water and the fish immediately decreases its respiratory movements (Belding 1929).

Under the exposure of ammonia, shrinkage, folding and loss of epithelium of the gill head and gill filament along with oozing of blood corpuscles were observed in the present Scanning Electron Microscopic Study. Besides these, some nodule-like structures were also seen on the surface of the gill head and gill filaments. Damages of similar nature of the gill head and filament surface have also been reported by earlier workers due to the effect of various pollutants. Rupture of the epithelium of the gill head and gill filament along with exposure of blood vessels was histologically shown by Christie and Battle (1963) in the gills of lamprey and trout at pH 5.3. Roy et al, (1986) under the long-term exposure of saponin on *Anabas testudineus* reported disassociation of epithelium from vascular layer both under histological as well as SEM studies. Ojha et al, (1989) observed lifting of epithelium from the basement membrane and its damage under Scanning Electron Microscope due to exposure of biocidal sap on the gills of the hill stream fish, *Garra lamta*. Munshi and Singh (1992), while studying the effect of low pH on the gills of *Channa punctatus* in scanning electron microscope, reported rupture of the epithelium of gill head and gill filaments and exposure of the blood vessels. Jha (1995) used a Scanning Electron Microscope to observe long term exposure of acid on *Labeo rohita* and viewed depressed on sunken gill head epithelium along with leakage of blood cells at gill filament surface. Cardoso et al, (1996) observed rupture, disassociation and disorganization of the epithelium of gill filament in *Lophiosilurus alexandri* under the treatment of unionized ammonia in histological preparation. In a Scanning Electron Microscopic study, Adhikari et al, (1997) reported shrunken epithelium of gill filament in *Heteropneustes fossilis* due to exposure to 15% seawater.

In another experiment, while working on the long term (20 days) treatment of salinity on the gills of *Labeo rohita*, Jha (1995) observed some puff or tumour-like swellings. Nodular or puff like structure in gill filament has also been reported by Roy et al, (1990) in *Heteropneustes fossilis*, Roy and Munshi (1991) in *Cirrhinus mrigala* and Sinha et al, (1997) in *Labeo bata*. It seems that the nodule-like structures observed in the present study are some outgrowth from the epithelium and may be considered similar to the puff or tumour-like structures described by others. The structures are probably the accumulation and proliferation of cartilaginous cells of the gill head (Sinha et al, 1997). Almost all the above-referred papers show damage and deformity of some kind or the other to the gill head and gill filament epithelium. Rupture and dissociation of the epithelium including oozing of blood cells etc. are the harmful effects of the contaminated environment. Such damaged epithelium fails to protect its surface from the adverse effect of the pollutants.

The increase in number and size of mucous glands and deposit of mucus makes a covering over the gills. Jozuka (1966), Narasimham and Parvatheswarao (1974); Agrawal et al, (1979); Roy et al, (1986); Roy and Munshi (1991) and Munshi and Singh (1992) also reported copious amount of mucus under the exposure of different pollutants. The increased mucus secretion is probably due to the proliferation of mucous cells under stressed conditions caused by the surrounding water (Perry and Laurent, 1989; Wipfli et al, 1994). In this concern, Mashiko and Jozuka (1962) explained that mucus reduces environmental osmotic pressure, whereas, Agrawal et al, (1979) suggested that changes in mucous cells are transient till the osmoregulatory ability is finally achieved. It seems that in the present study the copious amount of mucus over the gill surface is a response to protect it from the direct contact of polluted water. This seems reasonable as mucus is known to have some detoxification properties. While the deposit of mucus over the gill head beneficial as it protects the gill raker and gill arch epithelium from external contamination, increased mucus secretion over the lamellae adversely affects gaseous exchange due to increased water blood diffusion distance.

The loss of micro ridges observed in this study is due to the exposure to ammonia is a common phenomenon under the exposure of different types of pollutants or toxicants. A similar observation has been made by Roy et al, (1986) on the gills of *Anabas testudineus* due to the effect of saponin. They showed progressive loss of micro ridges leading to a frizzy appearance of micro ridged epithelial cells. Singh (1990) reported abbreviated lamellar surface due to breakage and some loss of micro ridges due to the treatment of heavy metals (copper and mercury) and low pH on the gills of *Channa punctatus*. Damage of microvilli and loss of cellular boundary of lamellae as observed in the present work resembles the work of Cardoso et al, (1996) on pacama (*Lophiosilurus alexandri*), in which they reported depression of plasmalemma in the lamellar cells under the treatment of ammonia. Damage of lamellar epithelium has also been observed by Daye and Garside (1976) at low pH (less than 5.6) in *Salvelinus fontinalis* but no remarkable change has been reported by Singh (1990) in acid (pH 5.3) exposed gills of *Channa punctatus*. It seems that the ammonia and not the pH is mainly responsible for the damage of cellular boundary and the micro ridges because no such changes have been reported by Singh (1990) even at pH 5.3. The actual mechanism of loss of micro ridges or microvilli is rather uncertain. It seems that the cellular swelling like a blown-up balloon may reduce the height of micro ridges and push them farther apart. Damages of lamellar epithelium result in the reduction of their surface area.

The fusion of adjacent secondary lamellae has been observed in the present experiment. Vander Pute and Part (1982) reported such fusion in the gills of rainbow trout after the exposure of 10.0mg.l^{-1} of hexavalent chromium at pH 6.5. Jacobs et al, (1981) observed corrugation and fusion of lamellae in *Salmo gairdneri* at increased experimental temperature. Lamellar fusion in *Channa punctatus* after exposure to copper and low pH was observed by Singh (1990). The occurrence of lamellar fusion has also been reported by Soderberg et al, (1984), Ojha et al, (1989) etc. The fusion of lamellae probably occurs due to intensified attraction between the epithelial cells of lamellae and leads to the formation of inter-lamellar bridges as observed during the long-term exposure of ammonia in the present study. The present observation corroborates with the earlier findings of Ojha et al, (1989) in *Garra lamta*; Singh and Munshi (1996) in *Channa punctatus*; Sinha et al, (1997) in *Heteropneustes fossilis* under the exposure of plant sap, malathion and 15% seawater respectively. Singh and Munshi (1996) explained that the gill epithelium influenced by water at acidic pH leads to dissociation and re-association of epithelium which finally results in the fusion of lamellae by an interlamellar bridge. It occurs due to certain injuries or infections. The fusion of lamellae and interlamellar bridge probably hamper the free movement of lamellae and reduces total respiratory surface area, therefore, it is not beneficial for gaseous exchange.

The presence of additional outgrowth of filaments filling up the interfilamental space and bifurcation of filaments at short-term exposure seems to be an adaptation to compensate for the insufficient oxygen uptake by the damaged lamellae and to cope up with the polluted environment. Burrows (1964), Reichanback-klinke (1967) also reported extra filamental growth during experimental conditions and thus showed similarity with the present observation.

Effect of ammonia on oxygen uptake: Respiratory metabolism of a fish varies with the changes in their environment as evidenced by the following noteworthy works of literature. Ranke–Rybicka (1975) observed a significant decrease in oxygen uptake when *Labistes reticulata* were exposed to different organophosphates. A significant decrease in oxygen uptake in malathion treated *Channa punctatus* was found by Pandey et al, (1976). Singh and Singh (1979) reported a considerable decrease in oxygen uptake in *Mystus vittatus* under different concentrations of copper and zinc. Shivraj and Patil (1985) observed an increase in oxygen uptake in *Lepidocephalichthys guntea* under the exposure of cadmium and copper. An increase in oxygen uptake in the fingerlings of *Labeo rohita* under short term exposure of alkaline pH and decrease during its long term exposure has been reported by Jha (1995). Smart

(1978) observed an increase in oxygen uptake in rainbow trout due to exposure to unionized ammonia.

In the present study, *Channa punctatus* showed an increase in the weight-specific oxygen uptake both at 24h (0.0435±0.0118mlO_2.g^{-1}.h^{-1}) and 48h (0.0703±0.0090 mlO_2.g^{-1}.h^{-1}) exposure of ammonia when compared with the control fish (0.192 ± 0.0040 mlO_2.g^{-1}. h^{-1}). However, it only increased to 0.0285±0.0086 mlO_2.g^{-1}.h^{-1} at 96h exposure. When compared to the oxygen uptake in controlled *Channa punctatus*, the increase at 24h, 48h and 96h exposure of ammonia was calculated to be 2.26, 3.36 and 1.48 folds respectively which are statistically significant (Table-22). While working with the acutely lethal effect of ammonia on rainbow trout (*Salmo gairdneri*), Smart (1978) found that oxygen uptake of the fish increased from a control value of 78.95±7.40 to 263.00±22.87 mlO_2.kg^{-1}.h^{-1} in an hour showing a 3.3 fold increase over the controlled value. It then declined to 194.80±30.80 mlO_2.kg^{-1}.h^{-1} in the second hour, showing only 2.4 fold increase. Singh (1990) found 34.44% increase in oxygen uptake during 24h exposure of 0.5ppm of $CuSO_4$ in *Channa punctatus* but a decrease of 16.66% (1.17 fold) occurred in 20 days exposure. In another experiment, she noticed 68.90% (1.80 fold) increase and 33.61% (1.58 fold) decrease in oxygen uptake under an above duration of 0.5 mg.l^{-1} of $HgCl_2$ in *Channa punctatus*. The lethal and sub-lehal treatment of commercial-grade organophosphorus (Disyston) showed an increase in oxygen uptake in *Colisa latia* during the first hour of treatment followed by a decrease in subsequent hours (Reddy et al, 1977). The above discussion suggests that the change in medium causes both an increase and decrease in oxygen uptake depending upon the duration of exposure and dose of pollutant, etc.

The comparatively high oxygen uptake of *Channa punctatus* following ammonia exposure indicates the stress imposed by the pollutant. The oxygen uptake obtained in this study confirms the suggestions made by Smart (1976, 1978) that underexposure of ammonia, the increase in oxygen uptake was related to an increase in opercular frequency due to increased energy required to maintain salt and water balance (Lloyd and Orr, 1969). It seems that the increase in oxygen uptake due to ammonia exposure despite the gill damage is due to the additional filamental growth and bifurcation of filaments, increasing gill surface area.

The opercular frequency of *Channa punctatus* showed an increase from 37.0±3.0 min^{-1} (during control) to 44.0±3.0, 50.0 ± 4.0 and 43.0±4.0 min^{-1} respectively at 24h, 48h and 96h exposure of ammonia(Table-22). Similarly, Smart (1978) found that the opercular frequency increased from 73.0±3.0 to 131.0±5.0 min^{-1} in the first hour and then to 115.0 ± 4.0 min-1 following exposure of ammonia. Singh (1990) reported that in *Channa punctatus*, the opercular frequency increases 52.0±1.0 min^{-1} (in control) to 58.0 ± 2.0 min^{-1} and 55.0±1.0 min^{-1} under the exposure of 0.5ppm of mercury at 24h and 120h respectively. It seems that the fish respond to a changing environment due to the toxicant by changing their opercular frequency (Belding, 1929). Anees (1975) found an increase in opercular frequency in *Channa punctatus* after the exposure of a higher concentration of malathion. However, Pandey et al, (1976) reported decreased opercular frequency in the same fish (*Channa punctatus*) under exposure to similar pollutant (i.e. malathion). In this experiment, also, an initial increase in opercular frequency has been observed with an increase in exposure hours and then a decrease occurred. Thus, the present work showed similarity with the observations of Smart (1978), Anees (1975) and Singh (1990). It is felt that the increase in opercular frequency is probably related to the increase in oxygen requirement due to increased salt excretion (Pandey, 1975) or increased activity of the fish due to irritation.

The effects of hyperammonemia on fish are not as extensively studied as is in mammals. However, studies done on fish have shown similar effects of high levels of ammonia as mammals. High ammonia concentration causes a respiratory problem, behavioral and physiological changes and can act as a stressor on fish which ultimately resulting in mortalities due to acute toxicity. Reduced opercular movement and later lethargic movements with the frequent surfacing of fishes were also due to severe respiratory stress due to ammonia toxicity in higher ammonia concentration.

Thus it may be concluded that damage to the gill, its epithelium and loss of micro ridges from the epithelial surface are some of the common alterations that take place under the short or long term exposure of not only ammonia but many other pollutants, The observations showed that such damages even if net severe, are sufficient to affect the normal functioning of the gills. It appears that excessive secretion of mucus, extra filament growth and the bifurcation of filaments etc. acts as defense responses to combat the deleterious effects of ammonia.

7

Evaluation of Suitable Packing Conditions

Introduction

In the past few decades, there has been a considerable demand for certain air-breathing fishes from India, mainly in the research laboratories of western countries. This is probably due to their unique modifications, especially in the respiratory and circulatory systems. Though, the air-breathing fishes are capable of extracting oxygen both from water as well as air, yet their transportation to long distances in living conditions posed several problems, such as hypoxia, injury, disease and finally death.

Proper sanitization of all the equipment and accessories is one of the essential steps for the study of the structure of gills, determination of oxygen uptake of fishes under normal conditions, under the association, under anaesthetization, under ammonia toxicity and all other steps of packing and transportation in this pandemic of COVID-19.

For effective live fish transportation, inside and out comprehension on the ideal ecological conditions for negligible pressure and greatest endurance is required. At the point when the fish leaves its normal climate, it should be in a self-supportable climate that provisions important necessities, for example, ideal temperature, oxygen and so forth.

Transportation of different life cycle stages (spawn, hatchlings, fry, fingerlings and adults) of fish is traditionally done in different types of containers. The age-old practice of transportation entails heavy mortality probably due to depletion of oxygen, hypertension of carbon dioxide, the toxicity of accumulating wastes, hyperactive strain and exhaustion, infection, injury and variation in pH, temperature, salinity and hardness etc (Jhingran, 1985). A portion of dissolved oxygen is utilized both by the fish and other microorganisms for their metabolism. As a result, the system becomes gradually deplete of oxygen and affects the fish. To overcome the problem of high mortality during transportation, it is essential to evaluate suitable conditions for their transportation. An inexpensive or less expensive method of live fish transport will help our aqua-culturists in many ways.

Fishes are transported live for many purposes viz., re-stocking, live marketing, or conveyed to measuring plants for the butcher. This transportation stage is definitive because of the work just as cost included and subsequently at most consideration is needed to forestall the item misfortune by mortality or injury as it impacts the overall revenue (www.aquastarmagazine.com, 2019).

Consideration must also be given to the space element. As far as, fry is concerned, the ratio between the volume of the fish transported and the transport water does not exceed 1:3. Heavier individuals, such as parent fish can be transported in a fish: water weight ratio of 1:2 to 1:3, although with smaller specimens, this ratio decreases to 1:100 to 1:200 (Pecha et al, 1983).

Maintenance of dissolved oxygen either through oxygenation or aeration of water is probably an important aspect on which the success of transport of live fishes largely depends (Saha et al, 1956a). In the open containers, aeration is done either by a regular change of water or by agitating the water with hand or pumping air into the container by an oxygen cylinder or a cycle pump (Selvaraj et al, 1981). However, in a closed condition, either the number of fishes should be kept as low as possible so that the oxygen content in water does not go below the required level or the hypoxic water is replaced by normoxic freshwater at short intervals. But, none of the above procedures are possible during long-distance transport taking more than 48 hours.

The closed systems are reflected by polyethylene bags and other sealed transport units. They are used mostly to transport of the early fry but also brood fish. The transport of fry in polyethylene bags with oxygen is especially widespread in the world, being used as a very effective method. It substantially decreases the overall volume and weight of transport water, enables public transport to be used for fish-transport purposes, makes it possible to extend the transport time, and is economically beneficial.

The methods of fish transport in a sealed medium are described in detail in several general studies (Orlov et al, 1974; Kozlov et al, 1977; Woynarowich and Horváth, 1980; Pecha et al, 1983) and in many separate specific studies (Bogdan, 1972; Snow et al, 1978; Ioshev, 1980; Varga, 1984; Pandit and Ghosh, 2012).

The polyethylene sac, or sleeve with a closed bottom end, is first placed in the outer transport case; if a double bag is to be used the inner bag must be inserted in the outer one. Then, water is poured into the bag, about 20 litres if the volume of the bag is 50 litres, and the fry is put inside.

The standards of fish fry transport published in the instructions by Orlov et al, (1974) for Soviet conditions can also be considered as practically verified.

The data apply to fish densities in bags, volume 40 litres, including 20 litres of water and 20 litres of oxygen, from the point of view of the possible total weight and number of fish transported; the standards are indicated for cyprinids, salmonids, perches and Soviet sturgeons.

The weight of fish that can be safely transported in a tank depends on the efficiency of the aeration system, duration of the transport, water temperature, fish size and fish species.

The ICAR-Central Institute on Postharvest Engineering and Technology, Ludhiana has built up a framework for transporting live fish to assist the farmers with selling their produce in live condition and acquire more pay (ICAR, 2019).

Mitra (1942) described the advantages of sending fingerlings of fish in a suitably designed container with oxygen. Raj and Cornelius (1949) transported 60 fingerlings of *Cyprinus carpio* from Ooty to Kumaon in sealed containers of 10 gallons capacity without any mortality. Ganapati and Chacko (1951) made a comparative study of the transport of fish seeds in ordinary tin containers with and without oxygen. Vass (1951) conducted a series of experiments by using sealed jerry cans to transport live fish under pressure and reported reduced mortality during the journey of 12 to 50 hours. Air transport of carp fry in packed plastic bags was put into practice in 1952 by the Department of Fisheries, Bombay, for the export of live tropical fish to England and other countries. Airlifting of carp fry is being carried out successfully by Central Inland Fisheries Research Station since the 1950s. Ranade and Kewalramani (1956) summarized that the air transport of carp fry is both simpler and cheaper than the rail transport method. Kulkarni and Ogale (1979) made air transport of mahaseer in India.

Six experiments were conducted by Aravindakshan et al, (2011) using table size fishes (425-1060 g) of three Indian major carps *viz., Catla catla, Labeo rohita, Cirrhinus mrigala* and the three exotic carps *Hypophthalmichthys molitrix* (Silver carp), *Cyprinus carpio* and *Puntius gonionotus*.

Various types of containers are used in different parts of India and abroad of which earthen vessels (=hundies), slots, special barrels, tarpaulin vats, cisterns, fish boxes, glass containers, open tin containers of various designs and closed containers with a supply of oxygen are common. State Fisheries Department of Orissa has improvised an open transport motor van with a small semi-rotatory pump. Watertight, tar-coated, plaited bamboo baskets (15 litre capacity) are used in Indonesia and reported to be cheaper (Hora and Pillay, 1962). In China, bamboo baskets and wooden tubs (2m high and 2m diameter) are reported to be in use. Department of Fisheries, Maharastra introduced container of polythene bags on an economical basis. Alkathene bags (74×46cm) are now widely used in India. Open canvas containers (1.0×1.0×2.5cm) and galvanized iron drums (180 litre capacity) are used in Madhya Pradesh and Punjab to transport major carps. Despite being cheaper, the open system of transportation is going out of use probably due to the involvement of constant care and frequent renewal of water in route. The transport of spawn and fry in oxygen sealed packs has gained popularity in India. In USSR, carp larvae (2000 to 4000 litre^{-1}) are transported in plastic bags having a 1:1 ratio of water and oxygen (Konradt, 1968).

Jagannadhan (1947), Basu (1951); Saha and Choudhary (1956), Hora and Pillay (1967) and Patro (1968) etc. have described different types of open systems for the transportation of live fishes. Transportation under a closed system had been described by Ranade and Kewalramani (1956); Mammen (1962a and 1962b); Patro (1968); Singh et al, (1979) etc. A standard open-live fish container is made up of galvanized tin with 40-50 litres capacity. It is mounted on a wooden base to prevent heating by contact with the ground. Khan (1939 and 1946) described a container (made up of galvanized iron, 45.72×35.36cm with two airtight openings) for transporting fish seed. Ganapati and Chacko (1951) made a comparative study of the transport of fish seed in ordinary tin carriers. CIFRI (the 1950s) introduced the use of kerosene tins filled with airtight screw-capped lids for transportation. Saha and Sen (1958) described containers of glass carboys, aluminum and galvanized iron sheets, latex rubber and plastic bags for long-distance transportation of spawn and fry under pressure and recommended that latex rubber bags are good for commercial use.

Live fishes are also transported without water in the crates of plywood (veneer), tin boxes or duralumin boxes by covering the bottom of the crate/box with moss, muslin pieces, mats of bast and reed. Privol'nev (1964) recommended that live fishes can also be transported in humid conditions, carrying fish in a state of anabiosis. Quite hardy fish (tench, crucian carp) can survive for a fairly long duration without water as they respire through the skin. The survival of fish without water is influenced by temperature, humidity, pO_2 and pCO_2 etc. Murrels due to possession of air-breathing organs may survive for a long time until the gill chamber is moist. Murrels are carried in pots with some water or in the basket with wet green weeds (Satynarayana, 1996).

The effect of carbon dioxide on oxygen uptake concerning the transportation of fishes has been studied by Basu (1950); Albaster and Herbert (1954) and Saha et al, (1956a). Khan (1940); Saha et al, (1956), Saha (1956) and Singh

(1977) worked out on the oxygen uptake by the fish for the effect of dissolved oxygen.

The addition of fine silt (red soil) coagulates the suspended organic pollutant and keeps down the contamination. Saha et al, (1956b) during the study of the transport of spawn and juveniles of major carps, used permutit (absorbs ammonia), activated charcoal (absorbs CO_2 and NH_3) amber lit IR, 100 (removes cations) and pulverized earth (as an absorbent to increase the survival period of the transporting fish.

Conditioning, an important aspect for transportation of fish has been reported by Jagannadhan (1947); Saha and Chowdhary (1956); Alikunhi (1957); Hora and Pillay (1962) and Srivastava and Karamchandani (1964) etc. Anaesthethics are used to reduce metabolic activities. Kewalramani and Gogate (1968) introduced a method of anaesthetization of carps by using Novocaine, Amobarbital sodium and Barbital sodium. Schoettger and Julin (1967) and Sylvester (1975) studied the response of MS-222 on salmonids and striped mullets. Jha (1995) evaluated the proper dose of Benzocaine for anaesthetization of *Catla catla* and *Labeo rohita*. The use of salt with anaesthetic is useful in preventing stress because salt reduces the strain of osmoregulation and controls osmotic pressure by making the water isotonic to blood (Gopalakrishnan, 1968).

Although attention has been given to improve transportation techniques, these improvements are mostly related to the transportation of carps. The transportation of air-breathing fishes is comparatively easier due to their bimodal oxygen uptake. However, information regarding their transport is lacking (Ghosh and Jha, 1994) and a lot has to be done to fulfill the ever-increasing demand for air-breathing fishes besides carps. Therefore, an attempt has been taken to investigate the suitable condition for long-distance transportation of live *Anabas testudineus* and *Channa punctatus*.

Materials and Methods

To develop suitable packing conditions for long-distance transportation of live *Anabas testudineus* and *Channa punctatus*, the following works were conducted namely:

(i) Estimation of oxygen requirement during transport,

(ii) Determination of a suitable dose of anaesthetics and

(iii) Experiments to evaluate the success of packing methods.

Live specimens of *Anabas testudineus* were procured from the North Bihar fish market and *Channa punctatus* was taken from the local fishermen. These fishes were kept in large plastic pools (1000 litre capacity) in the Departmental Aquarium House for about fortnight acclimatization. After weight wise segregation, these fishes were then transferred to the laboratory and maintained in plexiglass aquaria (60×30×30 cm) having chlorine-free tap water. The fishes were daily fed with chopped trash fishes and goat liver on alternate days.

The amount of oxygen required by *Anabas testudineus* and *Channa punctatus* of different body weights from aerial and aquatic routes were determined for transportation of different durations. This was estimated from the intercept and slope values of the oxygen uptake of these fishes as determined in chapter-III of this book and Ghosh (1984). Following the method of Martyshev (1983), the volume of water was calculated for a requisite hour of transportation of fishes of known bodyweight, firstly the oxygen requirement ($mgO_2.kg^{-1}.h^{-1}$) by the fish of 1.0kg body weight was calculated. The value was multiplied by a factor (Martyshev, 1983) for the desired temperature. This value was then divided by the amount of dissolved oxygen ($mg.l^{-1}$) present in water at a particular temperature (Privol'nev, 1964). The value from this division was again multiplied by the weight of fishes (kg) and duration of transportation (hours) to get the actual volume of required water (litre). The required volume of oxygen needed for aerial respiration was then filled in the poly bags using an oxygen cylinder.

Following the methods of Saha et al, (1956a), the weight and number of fishes consuming 1mg oxygen per hour were calculated. For this, the already determined values of oxygen uptake for the fish and their corresponding body weights were used. The average weight of fish(es) upon division with oxygen uptake ($mgO_2.h^{-1}$) gives the value of the weight of fish(es) to consume 1mg oxygen in one hour. This value was further divided with an average weight of the fish to get the number of fish that consume 1mg oxygen in one hour.

Suitable concentrations of different anaesthetics, namely Benzocaine and MS-222 were determined for proper sedation. For these 10 isomorphic fishes, each of either species (*Anabas testudineus* or *Channa punctatus*) was separately kept in small plexiglass aquaria (30×30×30cm) having a known volume of tap water. The experiments were repeated with the fishes of different body weights and the most suitable dose for I (2) sedation stage (Mc Farland, 1960) was determined. Sodium chloride and antibiotics were given following the doses as given by Gopalakrishanan (1968) with minor modifications.

Clearance concerning the fitness of the fishes was made available from a veterinary inspector or doctor. A no-objection certificate was obtained from the office of Marine Products Export Organization; Kolkata. *Anabas testudineus* has serrated operculum and spiny dorsal and pectoral fins, which generally damage the packing material during transportation. So, the serrated and spiny structures were trimmed out to prevent such damages. The trimmed fishes were treated with antibiotics and left for about a week for wound healing.

Before packing, methylene blue treated fishes were acclimatized for 4 to 5 hours, having fresh chlorine free water. Only healthy live fishes (i.e., fishes without injury, tumors and with intact scales and fins) were used for packing. Exhausted, dull and weak fishes were discarded for packing. Feeding was stopped before 24 hours of packing. For the packing experiment two sets of packs, one as control (i.e., without any treatment) and the other as experimental (i.e., with an anaesthetic, etc.) were made and observed for the desired period.

Polythene bags of different lengths and widths were used for packing. The bags were initially tested for leakage by placing each bag in an empty cardboard carton (having more height than the length and breadth). The flaccid bag when filled with water takes the shape of the container. Water temperature and oxygen content of water were recorded before introducing the fishes into poly bags. The same water was used for both sets.

In the bags treated as control, a known number of fishes of similar body weight were introduced, having a known volume of chlorine-free tap water. The bags were connected through the tube with an oxygen cylinder. The bags were squeezed shut by hand to the level of water surface to expel air and were slowly filled with the desired volume of oxygen from the oxygen cylinder. When the bags were filled with oxygen, it was squeezed shut by hand and then the upper part of the bag was folded over and tied with a rubber band. The readied bags were placed in a cardboard carton with the folded and tied end-directed upwards. Pieces of paper and ice were put between the poly bag and the container for insulation.

Under the experimental condition, an equal number of fishes of similar size were first treated separately with an antibiotic, Chloramphenicol, MAC Laboratories (P) Ltd. Bombay, at a dose of $10 mg.l^{-1}$. These bags were then filled with the same volume of chlorine-free tap water along with sodium chloride (500mg) and anaesthetic depending upon the size of the fish. The treated fishes were introduced in the bag and it was then filled with the required amount of commercial oxygen and packed. Many such bags were packed parallel to the control ones and kept for different hours.

The mortality of fishes and time were recorded for both control and experimental sets. Before introducing the packed fish in the stocking water, the packed plastic bags are floated on the stocking water for about 30 minutes, so that the temperature of the water-packed bag and that of the stocking water becomes the same. The bag was then opened carefully and then stocking water was slowly mixed with the water of the plastic bag and gradually the fishes were released into the stocking water.

Fishes once used for experiments on transportation purposes are further used for the same work only after a week because the fishes become sluggish during experimentation. A certificate was obtained from the consignor on reaching the destination, in which duration, condition of fish, loss (if any) with reasons was mentioned.

Observations and Results

The results of various experiments performed to determine the suitable packing methods are given below:

Determination of suitable dose of anaesthetics: The suitable concentrations for proper sedation (I[2] deep sedation level; Me Farland, 1960) of two anaesthetics namely Benzocaine and MS-222 for *Anabas testudineus* and *Channa punctatus* (bodyweight=1.0 to 17.0g) were determined and shown in Table-23. The anaesthetized fishes showed some loss of reactivity to external stimuli to vision, vibration and touch etc. These fishes became stationary in comparison to control fishes and settled down at the bottom of aquaria. They secrete a copious amount of mucus, which forms a layer over their body.

Anabas testudineus: An average dose of 20.875±6.277mg.l^{-1} of Benzocaine was found adequate to sedate *Anabas testudineus* of mean bodyweight 7.263±4.364g. It has been found that a dose of 14.0, 18.0, 23.0 and 28.50mg. l^{-1} of Benzocaine is suitable to sedate *Anabas testudineus* of 2.7±0.5, 5.1±0.9, 8.5±1.5 and 12.75±1.75g body weights respectively.

Similarly *Anabas testudineus* of body weights 4.45±1.25, 7, 30±0.70, 10.10 ± 0.90, 14.0±0.50 and 16.0±1.0g needs 40.0, 54.0, 62.0, 70.0 and 76.0 mg.l^{-1} of MS-222 respectively for sedation. Thus, an *Anabas testudineus* of average bodyweight 10.37±4.73g requires 60.40±14.10mg.l^{-1} of MS-222 for proper sedation (Table-23).

Channa punctatus: An individual of *Channa punctatus* having a bodyweight 6.775±4.231 g on an average requires 22.625±4.269mg.l^{-1} of Benzocaine for proper sedation. Similarly, *Channa punctatus* of body weights 2.05±0.85, 4.25±0.75, 7.40±2.20 and11.80±1.80 g need 17.0, 22.0, 24.5 and 27.0mg.l^{-1} of Benzocaine respectively as the suitable dose.

In another set of experiments, it was found that 20.0, 24.0, 27.0 and 32.5 $mg.l^{-1}$ of MS-222 were required respectively to anaesthetize 1.45±0.55g, 3.5±1.00g, 5.45 ± 0.55g and 11.55±1.65 g body weights of *Channa punctatus*. Thus, an average dose of 25.875±5.266$mg.l^{-1}$ of MS-222 is needed for proper sedation of a *Channa punctatus* of average bodyweight 5.488±4.359g (Table-23).

The average opercular frequency of a controlled *Anabas testudineus* was recorded to be 44.0±4.0 min^{-1} but it lowered to about 25.0±3.0 min^{-1} under sedation. In *Channa punctatus*, the average opercular frequency of 47.0±4.0 under controlled condition goes down to 28.0±3.0 min^{-1} under sedation.

Estimation of oxygen uptake: The oxygen requirement for transportation of *Anabas testudineus* and *Channa punctatus* has been calculated and summarized in Tables 24 and 25. The data were estimated from the values of bimodal oxygen uptake of these fishes as determined in chapter-III of this book. The data indicate that the rate of bimodal (aerial and aquatic) oxygen uptake increases with the increase in body weights.

Anabas testudineus: With the gradual increase in body weight from 2.70±0.50 to 16.00±0.60 g, the rate of aerial and aquatic oxygen uptake increased from 0.254 to 1.528 and 0.256 to 1.165 $mgO_2.h^{-1}$ respectively. The weight of fishes consuming 1.0 mg oxygen in one hour from aerial source showed a decreasing trend (10.630 to 10.471 g) but it had an increasing pattern (10.547 to 13.734 g) from aquatic source for the fishes of similar body weights. The number of fishes consuming 1.0 mg oxygen in one hour from air and water decreased gradually from 3.937 to 0.654 and 3.906 to 0.858 respectively, i.e., about from four fishes to one fish. From the values of aerial and aquatic oxygen uptake of *Anabas testudineus*, the oxygen required for transportation of five such fishes for 1h, 24h and 72h duration has also been calculated and shown in Table 24.

Table 23: Concentration of anaesthetics (Benzocaine and MS-222) for sedation of *Anabas testudineus* and *Channa punctatus* of different body weights.

Concentration of Benzocaine					
Anabas testudineus			*Channa punctatus*		
Body Weight (g)	**Total length (cm)**	**Anaesthetic concentration ($mg.l^{-1}$)**	**Body Weight (g)**	**Total length (cm)**	**Anaesthetic concentration ($mg.l^{-1}$)**
2.7±0.5	5.8 – 6.3	14.00	2.05±0.85	4.8 – 6.7	17.00
5.1±0.9	6.7 – 7.4	18.00	4.25±0.75	6.8 – 7.5	22.00
8.5±1.5	7.9 – 9.7	23.00	7.4±2.2	7.7 – 10.5	24.50
12.75±1.75	9.9 – 10.8	28.50	11.8±1.8	11.3 – 11.7	27.00
7.263±4.364		20.875±6.277	6.775±4.23		22.625±4.269

Concentration of MS-222					
Anabas testudineus			*Channa punctatus*		
Body Weight (g)	**Total length (cm)**	**Anaesthetic concentration (mg.l^{-1})**	**Body Weight (g)**	**Total length (cm)**	**Anaesthetic concentration (mg.l^{-1})**
4.45±1.25	5.9 – 7.2	40.00	1.45±0.55	4.2 – 5.8	20.00
7.30±0.7	7.0 – 8.3	54.00	3.5±1.00	6.2 – 7.0	24.00
10.10±0.9	8.0 – 8.9	62.00	5.45±0.55	7.4 – 8.1	27.00
14.0±0.5	9.4 – 10.0	70.00	11.55±1.65	11.3 – 11.5	32.50
16.0±1.0	10.2 – 10.9	76.00			
10.37±4.73		60.40±14.0	5.488±4.359		25.875±5.266

Channa punctatus: Aerial and aquatic oxygen uptake in *Channa punctatus* ranged from 0.095 to 0.705 and 0.172 to 1.048 $mgO_2.h^{-1}$ respectively with an increase in body weight from 1.45 ± 0.55 to 11.80 ± 1.80 g. The weight of fishes consuming 1 mg oxygen in one hour from air and water also gradually increased from 10.985 to 12.127 g and 8.430 to 11.259 g respectively with the above increase in body weight. The number of fishes consuming 1 mg oxygen in one hour from aerial and aquatic sources decreased from 7.576 (about 8) to 1.028 (about 1) and 5.814 (or 6) to 0.977 (or 1) respectively when body weight increased in a similar order. The oxygen requirement for the transportation of five specimens of *Channa punctatus* for 1h, 24h and 72h duration has been calculated from the values of their aerial and aquatic oxygen uptake and has been depicted in Table 25.

Estimation of water and oxygen requirement for transportation: Further, estimation was made to calculate the amount of oxygen required for 72-hour transportation of five individuals of *Anabas testudineus* and *Channa punctatus* of more or less similar body weight (7.35±0.05 g) from the calculated values. Thus, five individuals of *Anabas testudineus* (individual body weight: 7.30 g; total body weight: 0.0365kg) require 249.120 and 215.280 mg oxygen from aerial and aquatic sources respectively. Similarly, five individuals of *Channa punctatus* (individual body weight=7.40 g; total body weight= 0.037kg) will need 162.720 and 252.360 mg oxygen from air and water for transportation of 72 hours duration (Table-27).

Previously, it was calculated that at 22.0+1.0°C, *Anabas testudineus* and *Channa punctatus* of 1.0 kg body weight consumes 39.34 and 48.50 mg oxygen from water in one hour. The oxygen content of the water at 22.0°C is 8.73 ppm (Table-26). Thus, 39.34 mg oxygen will be present in 4.506 litre of water. Therefore, a total of 4.506 × 0.0365 × 72 = 11.840 litre of water will be required to transport five individuals of *Anabas testudineus* for 72 hours. Similarly, transportation of five specimens of *Channa punctatus*, a total of 5.556 × 0.037 × 72 = 14.800 litre of water will be needed because 48.50 mg oxygen is present in 5.556 litre of water. On the other hand, *Anabas testudineus* and *Channa punctatus* of 1.0 kg body weight need 94.79 and 61.08 mg oxygen from the air in one hour. Thus, a total amount of 94.79 × 0.0365 × 72 = 249.120 mg and 61.08 × 0.037 × 72 = 162.72 mg oxygen is needed from air for a 72 hour transportation of five individuals of *Anabas testudineus* and *Channa punctatus*. The commercial oxygen used for this purpose contains about 90% pure oxygen. Therefore, 277.78 mg (195.0 ml) and 181.0 mg (127.0 ml) commercial oxygen will be required.

Packing experiments: Using the above determined and estimated values, suitable methods for a closed system packing for long-distance transportation of live *Anabas testudineus* and *Channa punctatus* were made (Plates-X and XI). The dose of MS-222 for *Anabas testudineus* and *Channa punctatus* of body weight 14.50±0.5 g and 11.55±0.65 g was found to be 70.0 and 32.50 $mg.l^{-1}$ respectively (Table – 27). Five individuals of *Anabas testudineus* (total weight = 72.50g) need 385.880mg and 497.977 mg oxygen from aquatic and aerial sources respectively for transportation of 72.0 hours duration at 22.0^{0}C. Under similar conditions, the requirement of oxygen was 377.182 mg and 253.669 mg from water and air respectively for five specimens (bodyweight=57.75g) of *Channa punctatus* (Table–27). To obtain 385.880 and 377.182 mg oxygen from the aquatic source, 23.521 and 23.602 litre of water will be required.

For packing, polythene bags of 20 litre capacity were used. For transportation, the actual amount of water was decided considering the reduction in aquatic oxygen uptake due to grouping and use of anaesthetics. However, some additional oxygen was provided then the actual calculated amount of oxygen to cover the risk for safe transport. Based on the previous study, five individuals of *Anabas testudineus* (total weight=72.50g) were packed by giving 7.0g sodium chloride, 990.0 mg MS-222, 14.113 litre water (whereas the actual requirement of water was 23.521 litre) and 1.50 litre of commercial oxygen (instead of 497.977 mg or 348.0 ml required oxygen) in a polythene bag (Table–27). Similarly, 7.0g sodium chloride, 460.0 mg MS-222, 14.161 litre water (estimated need=23.602 litre) and 1.006 litre commercial oxygen (estimated requirement=265.669 mg or 185.969 ml) was used to pack five individuals of *Channa punctatus* (Table–27). The bags (consignments) were booked at 4.00 pm at Bhagalpur Railway Station for Kolkata by train. During the late-night of the next day, it was dispatched from Kolkata for South Bend, Indiana (USA) by an air–freighter service of KLM airlines. The consignment was received in Indiana after a total duration of about 72 hours without any mortality.

Table 24: Estimated values of oxygen required from aerial and aquatic sources for transportation of different durations of *Anabas testudineus* at 22.0±1.0°C. Values of weight and number of fishes consuming 1mgO_2.h^{-1} are also given

S. No.	Average wt. (g)	Oxygen uptake (mgO_2.h^{-1})		Weight of fish consuming 1mgO_2.h^{-1}		Number of fish consuming 1mgO_2.h^{-1}		Oxygen uptake mgO_2/5 fish/h		Oxygen uptake mgO_2/5 fish/72h	
		Aerial	Aquatic	Aerial	Aquatic	Aerial	Aquatic	Aerial	Aquatic	Aerial	Aquatic
1	2.70±0.50	0.254	0.256	10.630	10.547	3.937	3.906	1.270	1.280	91.440	92.160
2	4.45±1.25	0.420	0.392	10.595	11.352	2.381	2.551	2.100	1.960	151.200	141.120
3	5.10±0.90	0.482	0.440	10.581	11.591	2.075	2.273	2.410	2.200	173.520	158.400
4	7.30±0.70	0.692	0.598	10.549	12.207	1.445	1.672	3.460	2.990	249.120	215.280
5	8.50±1.50	0.807	0.681	10.533	12.482	1.239	1.468	4.035	3.405	290.520	245.160
6	10.10±0.90	0.960	0.788	10.521	12.817	1.042	1.269	4.800	3.940	345.600	283.680
7	12.75±1.75	1.215	0.961	10.493	13.267	0.820	1.041	6.075	4.805	437.400	345.960
8	14.00±0.50	1.335	1.041	10.487	13.449	0.749	0.961	6.675	5.205	480.600	374.760
9	16.00±0.60	1.528	1.165	10.471	13.734	0.654	0.858	7.640	5.825	550.080	419.400

Table 25: Estimated values of oxygen required from aerial and aquatic sources for transportation of different durations of *Channa punctatus* at 22±1°C. Values of weight and number of fishes consuming are also given

S. No.	Average wt. (g)	Oxygen uptake (mgO_2.h^{-1})		Weight of fish consuming 1mgO_2.h^{-1}		Number of fish consuming 1mgO_2.h^{-1}		Oxygen uptake mgO_2/5 fish/h		Oxygen uptake mgO_2/5 fish/72h	
		Aerial	Aquatic	Aerial	Aquatic	Aerial	Aquatic	Aerial	Aquatic	Aerial	Aquatic
1	1.45±0.55	0.095	0.172	10.985	8.430	7.576	5.814	0.475	0.860	34.200	61.920
2	2.05±0.85	0.132	0.232	11.202	8.836	5.464	4.310	0.650	1.160	46.800	83.520
3	3.50±1.00	0.221	0.366	11.475	9.563	3.279	2.732	1.105	1.830	79.560	131.760
4	4.25±0.75	0.266	0.433	11.580	9.815	2.725	2.309	1.330	2.165	95.760	155.880
5	5.45±0.55	0.337	0.538	11.695	10.130	2.146	1.859	1.685	2.690	121.320	193.680
6	7.40±2.20	0.452	0.701	11.859	10.556	1.603	1.426	2.260	3.505	162.720	252.360
7	11.80±1.80	0.705	1.048	12.127	11.259	1.028	0.977	3.525	5.240	253.800	377.280

Table 26: Amount of dissolved oxygen in the water at different temperatures (from T.I. Provl'nev)

Water temperature (°C)	Amount of dissolved oxygen ($mg.l^{-1}$)	Water temperature (°C)	Amount of dissolved oxygen ($mg.l^{-1}$)
0	14.56	16	9.85
1	14.16	17	9.65
2	13.78	18	9.45
3	13.42	19	9.26
4	13.06	20	9.09
5	12.73	21	8.90
6	12.41	22	8.73
7	12.11	23	8.58 (8.20)[1]
8	11.81	24	8.42
9	11.52	25	8.26
10	11.25	26	8.11
11	10.90	27	7.95
12	10.75	28	7.81
13	10.50	29	7.67
14	10.28	30	7.52 (7.68)[2]
15	10.06		(7.30)[3]

1 and 2 = FTR of ICAR, 1994 and 3 = Jha (1995)

Table 27: Details of different ingredients required for packing of five individuals of *Anabas testudineus* and *Channa punctatus* for duration of 72 hours.

No. of fishes	Name of fish	Total wt. (g)	Total bimodal oxygen uptake (mgO_2/5 fish/72h)		Amount of water given (litre)	Amount of oxygen given (litre)	Amount of anaesthetic (MS-222) ($mg.l^{-1}$)	Amount of salt (NaCl) ($mg.l^{-1}$)
			Aerial	Aquatic				
5	*Anabas testudineus*	72.50	497.977	385.880	14.113	1.500	70.00	500.00
5	*Channa punctatus*	57.50	253.669	377.182	14.161	1.006	32.50	500.00

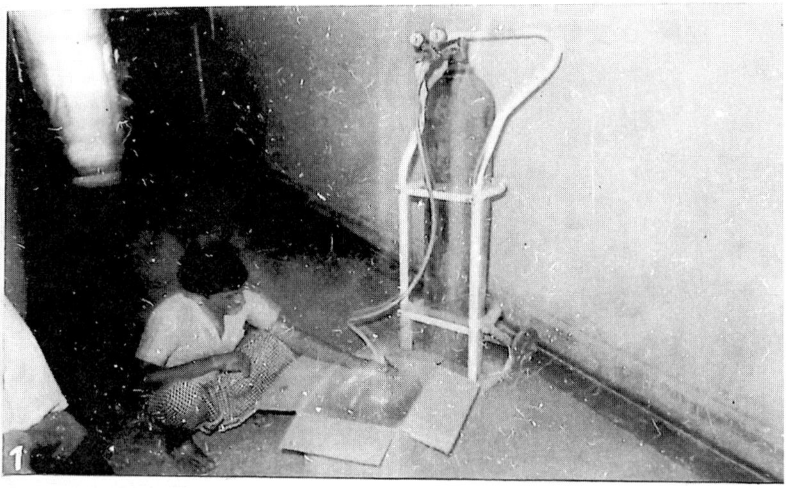

Explanation of Plate – X

1. Photograph showing filling of oxygen in the polythene bags containing fingerlings of an air breathing fish.
2. Photograph showing method of packing before filling of oxygen in the polythene bag.

Explanation of Plate – XI

1. Photograph showing a fisherman with an oxygen filled polythene bag containing known number of fingerlings of an air breathing fish.
2. Photograph showing different sets of polythene bags having fingerlings of an air breathing fish and oxygen in different ratio.

Discussion

Transport of different stages of live fishes has assumed considerable importance due to intensive fish-culture and the development of Inland Fisheries. Transport is necessary for breeding, stocking and marketing of fishes. Besides this, some air-breathing fishes are also used as a material for research work in western parts of the globe due to their various typical specializations owing to their habitat. Various Government Agencies of some countries from South East Asia are engaged in the artificial propagation of certain fishes and are the major transporters of live fishes. Fishes are transported either without any water (Martyshev, 1983) or with water using the tank method or plastic bag method. Different types of boats, steamers, ships, vehicles, and railways provide a means to transport live fishes among different states of a country. But airplanes are used for transporting fishes abroad in small quantities. The successful transport of live fishes needs tackling various problems such as

(a) Optimum number and size of fishes,
(b) Variation in temperature,
(c) Effect of handling,
(d) Dissipation of end products,
(e) Duration of transport and above all
(f) Requirement of oxygen.

The concentration of oxygen in the water and its uptake plays the most important role in the survival of fish during transportation. The oxygen concentration should be sufficient for normal respiration during transportation. It depends on the intensity of uptake by the fish during transportation as well as on water temperature. The oxygen content of water is affected by the pollution of organic substances excreted/secreted by the fish during transportation. The mucus secreted by the fish reduces the amount of oxygen in water by serving as a good substratum for putrefying bacteria (Martyshev, 1983). During a physicochemical analysis of water from different sources at Bhagalpur, it was observed that the amount of dissolved oxygen in the tap water was 8.20 ± 0.74 and 7.68 ± 0.48 ppm at a temperature of 23.0 ± 1.0°C and 30.2 ± 1.0°C respectively (FTR of ICAR, 1994). Similarly, Jha (1995) also reported a 7.3±0.1ppm of dissolved oxygen at 30.0±1.0°C (Table-26). The amount of dissolved oxygen in the water for transportation purposes was calculated to be 7.14±0.2 $mg.l^{-1}$ at 32°C by Singh (1977). Privol'nev (1964) tabulated the amount of dissolved oxygen in experimental water between 0°C to 30°C temperature (Table-26). He also calculated 8.58 and 7.52 $mg.l^{-1}$ of dissolved oxygen in water at a temperature of 23°C and 30°C respectively.

In the comparison of the values, it was found that the data are almost similar. Therefore, the Privol'nev table may be taken as a standard while considering the requirement of oxygen for transportation. The table showed that the amount of dissolved oxygen in water is reduced by 50% (14.56 to 7.52mg.l^{-1}) with an increase in temperature from 0°C to 30 °C. It indicates that at least two times more water will be required at 30°C in comparison to 0°C to fulfill the similar need for oxygen.

The mean oxygen consumption values in the range of 15.7-58.2 mg.fish^{-1}.h^{-1} and the computed oxygen consumption rates were 19.74-70.58 mg.kg^{-1}.h^{-1} during transportation in FRP tank was observed by Aravindakshan et al, (2011) for six live Asiatic carps namely *Catla catla, Cirrhinus mrigala, Labeo rohita, Cyprinus carpio, Hypophthalmichthys molitrix,* and *Puntius gonionotus*.

The rate of oxygen uptake of fishes during transportation varies considerably depending upon various biological (e.g., species, age, size, state of physical activity and feeding of fishes etc.) and abiological (such as pH, ammonia temperature, amount of dissolved oxygen and free carbon dioxide etc.) factors. In the present work, it was calculated that 72-hour transportation of five individuals of *Anabas testudineus* needs 497.977 and 385.880 mg oxygen from aerial and aquatic sources respectively at 22.0°C, whereas, the requirement was 253.669 mg (from the air) and 377.182 mg (from water) for *Channa punctatus* under similar conditions (Table-27). Privol'nev (1964) calculated that transportation of 250.0 kg of peeled (a type of white fish) at 10.0°C for 48-hour duration needs 1200000.0 mg oxygen from water. Singh (1977) observed that at 31.0 ± 0.5°C. 2712.0mg oxygen from water is required for 48 hours transportation of 50 individuals of *Cirrhinus miragala* (individual weight=6.4 g) and 9600.0mg oxygen by *Labeo rohita* (mean individual weight 14.5g) under similar conditions. However, the work of Privol'nev (1964) and Singh (1977) are related with similar purposes but are applied only for purely aquatic fishes. The account provides an idea that the difference in oxygen requirement in different fishes (when their number, weight and duration of transportation were kept similar) is probably due to differences in the behaviour of the fish at a particular temperature.

In the present study, it was calculated that 23.521 litre and 23.602 litre of water is required to get 385.880mg and 377.182mg oxygen respectively from aquatic source for a 72 hours transportation of five individuals of *Anabas testudineus* and *Channa punctatus* (Table-27). While working with the effect of anaesthetic on aquatic oxygen uptake, it was observed that *Channa punctatus* lowers its oxygen uptake by about 40% (chapter V of this book) due to the effect of Benzocaine. In another experiment (chapter IV of this book) it was calculated that *Anabas testudineus* and *Channa punctatus* showed at least 50% reduction in their aquatic oxygen uptake due to grouping under still water condition. Considering the lowering in oxygen uptake due to sedation and grouping, only 14.113 and 14.161 litre of water was given to the bags of *Anabas testudineus* and *Channa punctatus* instead of the actual requirement of 23.521 and 23.602 litre of water respectively. Jha (1995) observed a decrease of about 36.20% in the oxygen uptake of *Catla catla* and *Labeo rohita* under the treatment of Benzocaine. In another experiment, he also observed a lowering of 18.5 to 45.4% in the oxygen uptake of *Catla catla* under similar conditions. Motwani and Bose (1957), while establishing improved methods of transport of fry over long distances also discussed that the oxygen uptake of *Labeo rohita* shows dependency on the bodyweight of fishes irrespective of their number.

The requirement of oxygen from an aerial source needs special consideration due to the bimodal nature of the air-breathing fishes. Besides this, the utilization of some parts of oxygen for oxidation of metabolic products, by bacteria and other micro-organisms (Saha and Sen, 1958) from the ambient water is also important. In this experiment, 231.528mg and 226.300mg oxygen were available through the water in the bags of *Anabas testudineus* and *Channa punctatus* respectively against its actual requirement of 385.880mg and 377.182mg oxygen. It means that 154.352mg and 150.882mg oxygen from the aquatic source was reduced in the consignments of *Anabas testudineus* and *Channa punctatus* by reducing the amount of water. On the other hand, these fishes need 497.977mg and 253.669mg oxygen from aerial sources for a journey of 72-hour transport. As the amount of dissolved oxygen was reduced to make the consignment lighter or to lessen the weight of the consignment, the amount of aerial oxygen was increased proportionately. Considering the suggestions of Saha et al, (1956), an additional 25% oxygen as reserve oxygen was added to the bags. Singh (1977) found satisfactory results by enhancing an additional amount of half to one litre of oxygen during transport. However, Saha and Sen (1958) believed that a 20% reduction in the number of packed fishes will ensure safe transport. Thus, an amount of 2143 mg (1500ml) and 1437mg (1006ml) oxygen was calculated for the transportation of five individuals of *Anabas testudineus* and *Channa punctatus* respectively, since the purity of commercial oxygen gas was 90%, 2370mg oxygen (1666ml) and 1586mg oxygen (1118ml) of commercial oxygen was provided actually in the bags of *Anabas testudineus* and *Channa punctatus* respectively.

The use of the plastic bag method reduces the weight and chances of injury from hitting the wall of the container as against tin containers. For this purpose, polythene bags (made up of either polyethylene sheet, chlorvinyl sheet or polypropelyne sheet) of 10-20 litre capacity are used. These bags are transparent and offered facilities to observe the activities of fishes from outside.

Keeping the fish without food for at least 24 hours before packing is to prevent regurgitation of partially digested food into the water during transportation. Moreover, the requirement for oxygen is lower in starved fishes than in the fed ones.

It has been reported that for every millilitre of oxygen, a fish consumes, about 0.9 millilitre of carbon dioxide is released (Singh, 1977). The carbon dioxide is equally soluble in air and water therefore, once the water becomes hypercarbic, some part of carbon dioxide diffuses in the air. The aeration or oxygenation provides sufficient oxygen for uptake and keeps carbon dioxide at a low level. If the atmospheric oxygen diffuses by $1mg.h^{-1}$ then the corresponding rate of diffusion of pure oxygen under similar experimental conditions is $5mg.h^{-1}$ (Saha et al, 1956a). Therefore, at least five times more fishes may be transported in the same medium if atmospheric oxygen is replaced by pure oxygen. Mammen (1962a) suggested that in an open system the rate of diffusion of oxygen is about 20% from the atmosphere without any extra pressure, which improved to about 100% by providing surrounding oxygen under pressure in packed conditions. The oxygen above the water diffuses slowly due to the jolting movement under transit during transport. Therefore, transportation with diffused oxygen or compressed air is becoming more and more popular and is replacing the traditional methods.

Suggested method of packing for long-distance transportation: Values have been calculated for transportation of longer duration and also for more number of individuals of *Anabas testudineus* and *Channa punctatus*. During calculation, the number of individuals was kept constant (i.e., 5) when the duration of transportation was increased from 72 to 96 hour, on the other hand, when the duration was of 72 hours, the number of individuals was increased from 5 to 10. For this, many sets of polythene bags may be packed by reducing the volume of water and increasing the amount of oxygen to fulfill the deficiency of aquatic oxygen for 72-hour transportation of 10 individuals of *Anabas testudineus* and *Channa punctatus* and also a 96h transportation of 5 specimens of these fishes. During packing in sets II, III and IV the volume of water may be reduced by 25%, 50% and 75% respectively and the amount of oxygen to be obtained from the water was provided by adding extra oxygen than the required volume of oxygen.

For a 96 hour transportation of five specimens of each of *Anabas testudineus* and *Channa punctatus*, many sets of polythene bags may be packed by reducing the volume of water and increasing the amount of commercial oxygen (Table-28).

Anabas testudineus: In set I, 31.32 litre of water and 2.55 litre of commercial oxygen was given in the polythene bag, keeping the values of oxygen similar as it was calculated for water and aerial sources. 23.49 litre of water (25% reduction from the actual requirement) provided in set II and 2.64 litre of commercial oxygen was filled in the bags. In set III, 15.66 litre of water (50% reduced amount) and 2.87 litre of oxygen was supplied. Only 7.83 litre of water (a 75% reduced volume) was given in a bag as a set IV, in which 2.97 litre of commercial oxygen was blown through the cylinder (Table-28).

Channa punctatus: 31.47 litre of water and 1.667 litre of commercial oxygen was given in set I as it was calculated from the bimodal oxygen uptake. In set II, 23.60 litre of water (25% reduced amount) was poured carefully in the bag and then 1.807 litre of oxygen was supplied. 15.74 litre of water (50% reduction from the actual requirement) was provided in set III and 1.95 litre of commercial oxygen was filled in the bag. In set IV, only 7.87 litre of water (a 75% reduced volume) 2.09 litre of oxygen was provided as a source of aerial oxygen.

Following the similar trend in reduction in the amount of water, packing was make for a 72 hours transportation of 10 individuals of *Anabas testudineus* and *Channa punctatus*. The required volume of water and commercial oxygen for different sets (i.e., set I, II, III and IV) has been given in Table-28.

Out of different sets (i.e. I, II, III and IV), set IV seems more convenient because the size of the consignment was easy to carry due to its comparatively lightweight. About 300ml of commercial oxygen is sufficient to fulfill the requirement of aquatic oxygen from about 15.0 litre of water at 22°C.

8

General Discussion

In most of the underdeveloped and developing countries, regional diets are monotonous, unbalanced and poor in quality having a high percentage of starchy components and inadequate proportion of protein. The long-term solution to the problem of protein malnutrition lies in the greater production and better distribution of a protein-rich diet. Fish offers protein in a cheap and alternative form. Probably due to this, man has exploited the hydrosphere for food and consequently the history of the development of pisciculture. As the fish perishes rapidly, methods have been devised to keep the fish fresh between the catch and consumption.

The development of fisheries in India has attained considerable progress in increasing the per hectare yield of carps by controlled breeding and rearing of the young ones. Keeping the pace of development, the culture of air-breathing fishes has attracted the attention of aqua-culturists. Many varieties of these fishes have been successfully bred using pituitary extract (Ramaswami and Sundararaj, 1957; Moitra et al, 1979; Thakur, 1976). The traditional methods of transport of spawn, fry, fingerlings and even brooders in various open and closed systems pose major problems due to their heavy mortality. The mortality of live fishes is high mostly due to stress during transportation. For a long, studies on different aspects of transportation of carps and their allies have been carried out by fishery scientists. Some of the works which need special mention are that of Khan (1939, 1946), Ganapati and Chako (1951), Ranade and Kewalramani (1956), Saha and Chowdhury (1956), Saha et al, (1956a and b), Ramchandran (1969), Singh (1977), Selvaraj et al, (1981) and Jha (1995) etc. In comparison to Indian major carps, consumption of air-breathing fish is not so popular although they are a rich source of protein. However, their demand is gradually increasing. Much has to be done to meet the increasing demand of air-breathing fishes to exploit the maximum available derelict water.

The two commercially important Indian air-breathing fishes, *Anabas testudineus* (Bloch) and *Channa punctatus* (Bloch) commonly known as climbing perch and snake-headed murrel belong to the order Perciformes and Channiformes respectively. They live in different trophic zones of swamps and derelict ponds and are highly predatory habits. In the present work, oxygen uptake has been observed in *Anabas testudineus* and *Channa punctatus* concerning their body weight and group size. The effect of anaesthetic on the oxygen uptake and the effect of ammonia on the structure and function of the gills in *Channa punctatus* have also been studied. The findings have been used to evaluate suitable methods for long-distance transportation of these fishes. As we surviving in the COVID-19 pandemic, proper sanitization of all the equipment and accessories is one of the essential steps for fish transportation.

Oxygen Uptake Concerning Body Weight

Oxygen uptake in unit time is defined as metabolic rate and it represents the intensity of metabolism when related to unit body weight. The effect of body weight on the oxygen uptake in fishes is a subject of considerable discussion and has been extensively studied by various workers (Zeuthen, 1953; Parwatheswarao, 1960; Kamler, 1970; Munshi and Dube, 1973; Ojha and Munshi, 1975; Munshi et al, 1976; Ojha et al, 1977; Munshi et al, 1978; Biswas et al, 1979; Ghosh and Munshi, 1987; Ojha and Haque, 1994; etc.). It has been found that the air-breathing fishes show wide variation in their slope values for oxygen uptake under different conditions. Different slope values have been calculated for aquatic and bimodal oxygen uptake in these fishes. Information related to the effect of body weight on oxygen uptake in *Anabas testudineus* and *Channa punctatus* in adult fishes is available from the work of Ghosh and Munshi (1987) and Ghosh et al, (1990) respectively. Probably no attempt has been made to establish the relationship between oxygen uptake and body weight during the early developmental stages of these fishes.

Fish weight and water temperature are the main factors underlying oxygen uptake by fish concerning oxygen metabolism during transport. More oxygen is required for heavier fish transported in warmer water. For instance, if the water temperature increases by 10°C, oxygen uptake is about doubled. From the point of view of fish transport, for every 0.5°C rises in temperature, the fish load should be reduced by about 5.6%; conversely, for every 0.5°C decrease in temperature, the load can be increased by about 5.6% (Piper et al, 1982). Oxygen uptake also increases with fish excitement by handling. Excitement boosts oxygen demand three to five times and, for example, salmonid fry needs up to several hours to return to the normal level of oxygen metabolism which is, in fact, usually after the end of the transport (Lusk and Krcál, 1974).

In the present study, the slope for aquatic oxygen uptake ($\mu lO_2.h^{-1}$) was found to be 0.4522 in *Anabas testudineus* (body weight=9.56 to 834.83mg) and 0.4064 in *Channa punctatus* (body weight=8.22 to 925.50mg) (Table-7). But a slope of 0.526 and 0.629 respectively have been reported by Munshi and Dube (1973) in *Anabas testudineus* (body weight=8.0 to 65.0g) and by Hakim et al, (1983) in *Channa punctatus* (body weight=8.33 to 134.00g). These values are more than the values obtained in the present work. Munshi et al, (1978) reported that in *Heteropneustes fossilis*, the oxygen uptake through gills of juveniles (up to 25.0g) and adults (above 25.0g) increased by 0.478 and 0.799 respectively (Table-11). This indicates that the fishes showed intra-specific variation in aquatic oxygen uptake during different stages of their life cycle. The intra-specific differences in oxygen uptake are probably due to the effect of various activities during different stages of their life cycle, their different growth pattern, changes in feeding behaviour and physiological aspects of juveniles and adults (Smith, 1935a and b; Kamler, 1976).

As the slope was less than one, weight-specific aquatic oxygen uptake tends to fall with an increase in body weight. It decreased by a power of -0.5726 and -0.6141 respectively in *Anabas testudineus* and *Channa punctatus*. The decrease is mainly related to changes in skeletal structure and opportunities for low costs based on structural changes with size (Schmidt-Nielson, 1984). It is concluded that standard metabolic rate and the basal metabolic rate decreased by a power of 1 (isometrically) in larval fishes but less than 1 (allometrically) in juvenile and adult. Bennett (1988) emphasized that the allometric dependence of metabolic rate resides in the structural and functional organisation of the cell. Thus, the weight-specific oxygen uptake showed a negative slope also has been earlier reported by Pandey (1988); Singh et al, (1991) and Jha (1995) etc.

The regression analysis indicates that the slope of bimodal oxygen uptake during early developmental stages through air and water increased by 1.0019 and 0.3655 in *Anabas testudineus* and by 0.9533 and 0.3625 respectively in *Channa punctatus* (Table-8). In the adults, bimodal oxygen uptake from air and water in *Anabas testudineus* and *Channa punctatus* increased by a power of 1.009 and 0.851 (Ghosh and Munshi, 1987) and 1.067 and 0.8631 (Ghosh et al, 1990) respectively (Table-12). This shows that the slope for oxygen uptake from the air in these fishes is always higher (even more than 1.0) than the slope for oxygen uptake from water. The higher aerial slope value is perhaps related to the habitat of air-breathing fishes. These fishes survive in still water, where it has to breathe and re-breathe the same water by which partial pressure of carbon dioxide of water gradually increases which induce an increase in aerial breathing (Hughes and Singh, 1970a).

When aquatic oxygen uptake from continuously flowing and still water system was compared, it was found that oxygen uptake from still water gave lower slope values (0.3655 in *Anabas testudineus* and 0.3625 in *Channa punctatus*) than the continuously flowing water (0.4522 in *Anabas testudineus* and 0.4064 in *Channa punctatus*) (Table-7 and 8). Ojha et al, (1978) also reported a lower slope (b=0.683) from still water than continuously flowing water (b=0.765).

The present work reveals that *Anabas testudineus* and *Channa punctatus* depend more on water (60% and 68% respectively) for their metabolic activities during their early developmental stages. But, they chiefly adopt air-breathing as they grow in size. Ghosh and Munshi (1987) and Ghosh et al, (1990) reported that *Anabas testudineus* (body weight= 48.71g) and *Channa punctatus* (body weight=48.80g) consumed about 55.105±0.545% oxygen from the air. In this work, the quotient between aquatic and aerial oxygen uptake in the early developmental stages of *Anabas testudineus* and *Channa punctatus* was found to be 1.827 and 2.612 respectively which decreased to 0.814 in *Anabas testudineus* (Ghosh and Munshi, 1987) and 0.834 in *Channa punctatus* (Ghosh et al, 1990) during their adult stages. The decrease in quotient also indicates the dominance of aquatic breathing over aerial respiration during the early developmental stages. Therefore, it seems that the efficiency of oxygen uptake of the gills and accessory organs varies mainly with the growth of the fish.

It is evident from the data of oxygen uptake of adult *Anabas testudineus* and *Channa punctatus* that they mostly depend on aerial respiration to meet their total metabolic demand as their body weight increases. However, juveniles of these fishes were found to be more efficient in aquatic breathing. The diffusing capacity of gills decreases with the increase in the weight of fishes. The idea is supported by the works of Munshi et al, (1976) on *Clarias batrachus* and Munshi et al, (1977) on *Heteropneustes fossilis*.

The present observation also clearly indicates that the early developmental stages of *Anabas testudineus* would require more oxygen in the transport medium than those of *Channa punctatus* of the similar body weights. This information is important from the viewpoints of packing for transportation. Once the rate of oxygen uptake by a fry or a fingerling of a particular size is known, the number that can be transported is easily determined in a given volume of water.

Effect of Group Size on Oxygen Uptake

Works on fish aggregation reveal that oxygen uptake of a fish is changed by the number of fishes mutually participating in respiration and this has been demonstrated by various workers such as Allee (1931), Schtefanesku (1958), Grigor'eva (1966), Wares II and Igram (1979), Imabayashi and Takahashi (1987), Ghosh et al, (1994) and Jha (1995) etc.

The present study showed that the rate of aquatic oxygen uptake of *Anabas testudineus* and *Channa punctatus* goes down with an increase in the number of individuals in the respirometer either in continuously flowing water or in still water condition. In *Anabas testudineus*, with the successive increase in-group size from 5 to 40 fishes, the weight-specific oxygen uptake decreased from 0.140±0.026 to 0.080±0.006 $mlO_2.g^{-1}.h^{-1}$ (Table-13). The reduction in oxygen uptake as compared to the solitary fish was found to be significant ($p<0.01$, Table-17) when the fishes are in a group of 20 and onwards. Similarly, the weight-specific oxygen uptake lowered from 0.361±0.104 to 0.174±0.083 $mlO_2.g^{-1}.h^{-1}$ in *Channa punctatus* with a gradual increase in group size from 5 to 40 fishes (Table-14). A highly significant correlation ($p<0.001$) was found between group size and aquatic oxygen uptake (Table-17), The reduction in oxygen uptake under continuously flowing water system was calculated to be 27.08 to 58.39% (46.03±15.61%) in *Anabas testudineus* but the decrease was 28.55 to 65.52% (50.66±11.88%) in *Channa punctatus* when the number of fishes gradually increased.

In the still water also, the rate of aquatic oxygen uptake has been found to decrease with an increase in group size, Here also with the gradual increase in group size (5 to 30), the weight-specific oxygen uptake of *Anabas testudineus* decreased significantly ($p<0.001$, Table-17) from 0.053±0.070 to 0.012±0.003 $mlO_2.g^{-1}.h^{-1}$ (Table-15). Similarly, in *Channa punctatus* the oxygen uptake in still water decreased from 0.820±0.180 to 0.072±0.030 $mlO_2.g^{-1}.h^{-1}$ (Table-16) showing a highly significant correlation ($p<0.001$, Table-17). However, in still water, the oxygen uptake was always found to be higher than the controlled (i.e. continuous flowing) system. When compared to the controlled condition, the increase in oxygen uptake under still water was 114.72±39.20% in *Anabas testudineus* while it was 80.49±19.65% in *Channa punctatus* under similar conditions. The higher rate of oxygen uptake in still water conditions was probably related to the stressful condition of fishes.

These results showed that both *Anabas testudineus* and *Channa punctatus* follow similar trends in their nature of oxygen uptake. It may be concluded that the rate of aquatic oxygen uptake reduced with an increase in group size, both in continuously flowing water as well as in still water.

Earlier workers observed that the rate of oxygen uptake in fishes due to companionship or group effect is either decreased (Allee, 1931; Shlaifer, 1938; Job, 1955; Delco, 1963; Uematsu, 1971; Itazawa et al, 1978; Jha, 1995 etc.) or increased (Bowen, 1932; Schuett, 1934; Brett and Sutherland, 1965; Umezawa and Kanikawa, 1980; Ghosh et al, 1994; etc.). They have attributed various reasons for the variation in oxygen uptake by a fish in a group. It seems that oxygen uptake in fishes under grouping is probably related to their behaviour. The fishes become restless and excited when away from the association of intra-specific isomorphic members. Shlaifer (1938), while working with goldfish, discussed that the soothing effect reduces the excitability of fish and consequently the oxygen uptake in an association. Itazawa et al, (1978) explained that fishes are more timid and nervous under isolation or in a small group and therefore, have higher oxygen uptake. Imabayashi and Takahashi (1987) suggested that a group effect on oxygen uptake is correlated with the changes in behaviour with growth and development due to mutual contribution.

The juveniles of *Anabas testudineus* and *Channa punctatus* showed different behaviour in different conditions. They swim separately and randomly in a large area but prefer to remain in a shoal. A shoal is a group of fish for some social functions. As the *Anabas testudineus* (body weight=0.6 to 5.0g) and *Channa punctatus* (body weight=0.7 to 1.6g) did not show schooling behaviour, the reduction in oxygen uptake of these fishes in a grouped conditions is probably a result of such shoaling behaviour because they feel at ease in a shoal. Therefore, it may be concluded that the decrease in oxygen uptake in these fishes is due to the shoaling or soothing effect.

The work concludes that the early developmental stages of *Anabas testudineus* and *Channa punctatus* may be transported in a larger number (30 or 40) by reducing their oxygen uptake by about 25% because at least 25% reduction in oxygen uptake was observed in these fishes when they are in a group both in continuously flowing and still water conditions during experimentation.

Effect of Anaesthetic on Oxygen Uptake

During the transport of fish seed, fry and adult fishes, anaesthetics are used to reduce the stress or excitability, the metabolic rates, the release of carbon dioxide and waste products. Thus the chances of injury are also reduced. Due to the above benefits, the use of anaesthetics has earned considerable importance in fishery science, particularly during the transportation of fishes. This has led to the frequent use of anaesthetics in large quantities with inadequate knowledge of long-term effects either on its user or the fish itself. In India, anaesthetics are used mainly for the transportation of fishes since 1960. In practice, brooders are usually sedated either by dissolving the anaesthetic in water, spray of anaesthetic on the fish body or injecting the drug to the fish body. Jhingran and Pullin (1988) stated that anaesthetization of brood fishes is essential before transportation but is not necessary for fingerlings.

It is safest to sedate the fish in the holding facility for 30 minutes before loading and then to begin exposure to a lower concentration of sedative during transport. The use of anaesthetics should not be relied upon for increased load-carrying capacity.

Fereira et al, (1984) recommend transporting *Oreochromis mossambicus* treated with benzocaine-hydrochloride at a dose of 25mg.l-1. A slow-reacting, long-lasting tranquilizing effect on trout is developed also using sodium amytal, one of the several hypnotic barbiturates available. To some extent, the efficacy of the drug is controlled by temperature. According to Leitritz and Lewis (1976), it tends to decrease when the temperature increases above 12°C, the most effective range being from 8 to 12°C at a concentration of 7mg.l^{-1}. The effective use of sodium amytal in conjuction with barbital preparations is also reflected by Strebkova (1971). Carbonic acid might also be used for tranquilizing fish. Dupree and Huner (1984) recommend the concentration of 0.1–0.4 mg.l^{-1}, Mishra et al, (1983) transported fry (0.8 g) of *Labeo rohita* under the concentration of 0.5 ml.l^{-1} of carbonic acid. The product called Combelen (Bayer) was found to be effective in the transport of trout; this chemical is a neuroleptic which does not induce direct narcosis, but markedly reduces the effects of stress on the fish during transport (Studnicka et al, 1982); for trout at a weight of 250–300 g kept in 5°C water, a Combelen concentration of 0.2 mg.l^{-1} was found to be effective.

Since 1966, Novocaine has been used to sedate major carps for transportation purposes by the Department of Fisheries, Bombay. Kewalramani and Gogate (1968) used Novocaine, Amobarbital sodium and Barbital sodium to sedate certain Indian major carps. MS-222, although a registered fish anaesthetic in the United States, is not highly recommended for commercial transportation, probably due to higher cost, its toxic nature in seawater after exposure to light, hyperglycemia, damage to olfactory epithelium etc. Benzocaine includes ease of preparation, long storage life and safety. Ross and Ross (1984) have used Benzocaine (dose 33 to 200 mg.l^{-1}) to sedate salmonids, tilapia and catfishes.

Jiandong et al, (2010) studied the anesthesia of juvenile *Oreochromis niloticus*. When the dose of MS-222 was 10mg.l^{-1}, 20mg.l^{-1} and 30mg.l^{-1}, the average oxygen uptake decreased by 10%, 31.4% and 40.5%, and the average ammonia excretion rate decreased by 12.4%, 37.8%, 42.8%. When the dose of MS-222 was 20-100mg.l^{-1}, the oxygen uptake of *Procambarus clarkia* could not be reduced but decreased when the dose exceeded 100 mg.l^{-1}. The general trend was that with the increase of the concentration, the oxygen uptake and the rate of ammonia excretion increased first and then decreased.

In the present study, the effect of Benzocaine on the oxygen uptake and opercular frequency has been observed in the individuals of *Channa punctatus*. A dose of 18.0±3.0 $mg.h^{-1}$ of Benzocaine was found suitable for their sedation. A controlled juvenile of 3.71±1.78g body weight consumed 0.097±0.019 $mlO_2.g^{-1}.h^{-1}$, whereas, 3h and 24h Benzocaine exposed fish of the same bodyweight consumed 0.056±0.009 and 0.059±0.010 $mlO_2.g^{-1}.h^{-1}$. As compared to the control, the 3h and 24h Benzocaine exposed *Channa punctatus* showed 42.43±3.03% and 39.14±2.09% reduction in their oxygen uptake (Table-19). The average opercular frequency of 47.0±2.0 min-1 under the controlled conditions of a juvenile of above body weight reduced to 27.0±2.0 in 3h and 29.0±2.0 in 24h exposed fish. The opercular frequency thus showed about 42.81±3.21% and 39.68±2.95% reduction in 3h and 24h exposure (Table-20). The reductions both in oxygen uptake and opercular frequency in comparison to the controlled one were found to be highly significant ($p<0.001$) and showed a similar pattern. It suggests that the reduction in oxygen uptake, as well as opercular frequency by the juveniles of *Channa punctatus* in 24h exposure, is less pronounced than the 3h exposed fishes. This is probably due to some acclimatization at 24h exposure time.

The reduction in oxygen uptake was probably due to the lowering of metabolism, reduced opercular movement under sedation of Benzocaine. It was found that the fish secrete enormous mucus, which forms a film or coat over the gills. This enhances the water/blood diffusion barrier and may also reduce oxygen uptake.

Different fishes respond differently to different anesthetics, though MS-222 does not affect the crayfish. In general, for the same species with different requirements, the higher the quality is, the more the anesthetic dose is required. As with the same type and different ages of fish, because of biochemical reasons, respiratory organs are different; the effective anesthetic dose also has obvious changes. In the juvenile stage, skin breathing is the main method, while in the adult stage; gill breathing is the main method. As the age increases, the tolerance to anesthetics will lead to different effective anesthetic doses (Xiaoqin et al, 2017).

In the aquatic environment where fish grow, when the water temperature increasess but the dose of anesthetics is the same, the time for the receptor to enter anesthesia stage becomes shorter. When the water temperature increases to a certain temperature, the anesthetic effect of some anesthetics will significantly decrease. In most cases, the survival rate of recipients after anesthesia is greatly reduced at higher temperatures.

The change of pH has different effects on the fish body due to various doses. Generally speaking, it has little impact on freshwater fish and a strong impact on seawater fish. When high-dose MS-222 is used, a certain amount of $NaHCO_3$ should be applied to buffer it so that the pH is between 7.0 and 7.5 and lower the stimulation to the fish (Chun et al, 2016).

The dissolved oxygen has a profound effect on the respiration of fish. When dissolved oxygen is low, fish anesthetics play a faster role, but at the same time, recovery time is slower. When dissolved oxygen is high, the receptor is more sensitive to fishery anesthetics, and the dissolved oxygen the receptor needs is low, and the coming of anaesthesia stage of the receptor is slower, but the recovery time is shorter and the survival rate is higher.

Since the present observation clearly showed that Benzocaine brings about a considerable reduction in oxygen uptake, more number of fry and fingerlings may be transported in a given volume of water. Use of Benzocaine in place of MS-222 may be recommended because the former is economically less expensive and a lower dose of Benzocaine brings about the required sedative effect on the fish. Therefore, the given concentration of Benzocaine may be used safely for transportation of *Channa punctatus*.

Effect of Ammonia on Gill and Oxygen Uptake

Ammonia is the main excretory product in fishes, which is synthesized in the liver and passed through the gills. Its accumulation in the fish body induces alternations and may cause the death of the fish. Acute or chronic and lethal or sub-lethal exposure of ammonia on fishes has been critically studied by Brockway (1950), Llyod (1961); Robinette (1976); Smart (1976); Thurston et al, (1983) and Linton-Tyler et al, (1998) etc. The effects of ammonia exposure on the gill surface ultra-structure have been observed in the present study. In addition to this efficacy of the gills i.e. oxygen uptake was also observed due to exposure to ammonia both at pH 7.7 and 6.5.

Exposure of ammonia at 24 hours showed wrinkling of the epithelium in the gill arch and gill raker, folding of the epithelium at the base of the gill arch, numerous mucous gland openings and micro-ridged epithelial cells with the frizzy appearance of micro-ridges (Plates-IV and V), But, at 20 days exposure, it was found that opening of mucous glands become minute and the epithelium showed broken micro-ridges. The changes corroborate with the earlier findings of Roy et al, (1986); Cardoso et al, (1996); Adhikari et al, (1997) etc. under the exposure of different pollutants.

Exposure of acidic solution of ammonia for different durations showed multifold changes. At 24 hours of exposure, adjacent filaments showed shrunken epithelium and few bifurcated filaments; lamellae showed fusion at the tips of filaments, partial loss or even damage of the epithelium; invagination occurred in micro ridged epithelial cells. Numerous mucous gland openings and micro ridged epithelial cells along with the deposit of mucus at the filamental surface were observed at 48 hours of exposure (Plates-VI and VII). During long-term exposure, the rupture was observed in the epithelium; few lamellae of adjacent filaments formed interlamellar bridges (Plates-VIII and IX). The increase in number and size of mucous glands and deposition of mucus has also been reported by Jozuka (1964); Agrawal et al, (1979); Munshi and Munshi (1986) and Munshi and Singh (1992) due to exposure to different pollutants. The fusion of adjacent secondary lamellae was also reported by Vander Putte and Part (1982) and Singh (1990) etc. Dissociation and re-association of epithelium resulting in the formation of interlamellar bridge showed similarity with the work of Singh and Munshi (1996). The appearance of bifurcation of filaments is in the line of work of Burrow (1964) and Reichanback-Klinke (1967).

Following changes were noticed during different hours of exposure to an alkaline ammonia solution. At 24 hours of exposure, folding of the epithelium at the base of gill filament was more conspicuous; an additional outgrowth was seen in the filaments; lamellae were present up to the tip regions and their surface lost definite cell boundaries; micro ridges are broken into villi (Plates-IV and V). At 48 hours of exposure, the gill arch surface showed damage and loss of epithelium, exposure of erythrocytes, rupture of few mucous gland openings and loss of micro ridges of some micro-ridged epithelial cells and complete fusion of lamellae at the filament tips. Some nodular or bleb like structure were seen after long term exposure to an alkaline solution of ammonia. Rupture of epithelium and exposure of blood cells was also shown by Christle and Battle (1963), Munshi and Singh (1992) and Jha (1995) due to different pollutants. Damage of micro-villi and loss of cellular boundary in the present observation is in line with the work of Cardoso, et al, (1996). Observation of nodule-like structures was earlier reported by Roy et al, (1990), Jha (1995) and Sinha et al, (1997) under the exposure of saponin, saline water and *Strychnous nux-vomica* respectively.

Exposure to ammonia also affects the opercular frequency and consequently the oxygen uptake (Smart, 1978). In the present work, it was observed that the opercular frequency of 37.0±3.0min^{-1} in a controlled *Channa punctatus* increased significantly to 44.0±3.0min^{-1} at 24h and 50.0±4.0min^{-1} at 48h exposure but lowered to 43.0±4.0min-1 at 96h exposure of ammonia (Table-22). It showed 18.92%, 35.14% and 16.22% increase in opercular frequency when compared to the opercular frequency of the control fish during 24h, 48h and 96h of treatment respectively. Smart (1978) also found that opercular frequency increased from 73.0±3.0min^{-1} to 131.0±5.0 and then to 115.0±4.0min^{-1} due to exposure to ammonia in rainbow trout (*Salmo gairdneri*).

The weight-specific oxygen uptake of a controlled *Channa punctatus* (body weight= 7.176±0.491g) was calculated to be 0.0192±0.0040 $mlO_2.g^{-1}.h^{-1}$ but it changed significantly to 0.0435±0.0118, 0.0703±0.0090 and 0.0285±0.0086 $mlO_2.g^{-1}.h^{-1}$ respectively during 24h, 48h and 96h of exposure of ammonia to show 119.23%, 278.40% and 48.86% increase (Table-22). While working with the toxic effect of ammonia on *Salmo gairdneri*, Smart (1978) found that oxygen uptake of fish increased from a control value of 78.95±7.40 to 263.00±22.87 $mlO_2.kg^{-1}.h^{-1}$ and then declined to 194.80±30.80 $mlO_2.kg^{-1}.h^{-1}$. Various theories have been proposed to explain the reasons for either increase or decrease in opercular frequency and oxygen uptake under the influence of a pollutant (Smart, 1978; Shivraj and Patil, 1985). In the present study, the higher oxygen uptake of *Channa punctatus* may be due to the stress imposed by the pollutant, formation of additional filamental growth, bifurcation of filaments etc. which results in the increase of respiratory surface area and consequently the oxygen uptake.

Evaluation of Suitable Packing Conditions

Although the air-breathing fishes are capable of extracting oxygen both from aerial and aquatic sources yet their long-distance transportation in live conditions poses several problems such as hypoxia, injury, disease and finally death. However, transport of different stages of these fishes assumed considerable importance both for consumption as well as for research. The fishes are generally transported to a short distance by the use of earthen containers and to long-distance using plastic bag methods.

The dissolved oxygen in the supply water was found to be 8.20mg.l^{-1} and 7.68mg.l^{-1} at 23°C and 30°C respectively. Privol'nev (1964) tabulated the amount of dissolved oxygen to be 8.58mg.l^{-1} at 23°C and Jha (1995) determined dissolved oxygen of supply water to be 7.3mg.l^{-1} at 30°C (Table-26). These works conclude that the amount of dissolved oxygen mainly depends on water temperature, which affects the metabolic activities of the organisms present in the water.

During the determination of a suitable dose of anaesthetics, it was found that 20.875 ±6.277 mg.l^{-1} of Benzocaine is adequate to anaesthetize *Anabas testudineus* of body weight 7.263±4.364g. Similarly, an *Anabas testudineus* with a mean weight of 10.37±4.73g requires 60.40±14.10mg.l^{-1} of MS-222 for proper sedation. On the other hand, a *Channa punctatus* having body-weight 6.775±4.231g needs 22.625±4.269mg.l^{-1} of Benzocaine for the same purpose and an average dose of 25.875±5.266mg of MS-222 is needed for proper sedation of a *Channa punctatus* of mean body weight 5.488±4.359g (Table-23). For experimental purposes, a dose of MS-222 for *Anabas testudineus* (14.50g) and *Channa punctatus* (11.55g) was determined, which was 70.0mg.l^{-1} and 32.50mg.l^{-1} respectively. Ghosh and Jha (1994) also determined a sedative dose of MS-222 to be 85.0±1.5 mg.l^{-1} for *Anabas testudineus* (bodyweight 14.5±1.5g) and only 38.0±11.5mg.l^{-1} for *Channa punctatus* (bodyweight 36.5±1.5g).

Singh (1977) observed that 2712.0mg oxygen from water is required for the duration of 48h transportation of fifty individuals of *Cirrhimus mrigala* (average body weight=6.4g) and 9600.0mg oxygen by *Labeo rohita* (average body weight=14.5g). Based on the available data on the oxygen uptake of *Anabas testudineus* and *Channa punctatus* and the work of Singh (1977), it was calculated that five individuals of *Anabas testudineus* need 497.977 and 385.880 mg oxygen respectively from aerial and aquatic sources for 72h transportation (Table-27). For the same purpose, the requirement of oxygen was found to be 253.669mg from the air and 377.182mg from water for five specimens of *Channa punctatus*. During packing, 14.113 and 14.161 litre of water was provided respectively to fulfill the requirement of aquatic oxygen by these fishes. The reduced amount of water was given considering the reduction of aquatic oxygen uptake due to group size and anaesthetic.

Similarly, 1.500 and 1.006 litre of commercial oxygen was given to the bags of *Anabas testudineus* and *Channa punctatus* respectively taking the factors responsible for safe transport. Such calculations are probably not reported earlier and were calculated based on the methods used by Privol'nev (1964) for the requirement of oxygen from water.

Based on these works, the oxygen requirement of fishes concerning their transportation may be calculated applying the relation aW^b, duration and number of individuals to be transported.

9

Summary and Conclusion

Fish transport is a vast area comprising the problems of purely technical design on the one hand, and the chemistry of water, biological reactions of fish and the like, on the other. The above survey could only briefly mention and evaluate the relevant literature on these problems. Because of its purpose, the book was designed not only as an analytical-synthetic survey of pertinent literature but, at the same time, as a practical manual to be used for seeking at least partial answers to practical questions concerning fish transport. The live fish trade is a rapidly expanding component of commercial fishing and aquaculture industries.

Proper sanitization of all the equipment and accessories is one of the essential steps for the determination of oxygen uptake of fishes as well as all the other steps of fish transportation in this pandemic of COVID-19.

From this work, the following conclusions were drawn:

1. During the early developmental stages of *Anabas testudineus* (body weight=9.56±4.88 to 834.83±5.30mg), the aquatic oxygen uptake per unit time gradually increased from 25.216±6.711 to 251.444±9.655$\mu lO_2.h^{-1}$ while the corresponding weight-specific oxygen uptake decreased from 3.087±1.132 to 0.301±0.009 $\mu lO_2.mg^{-1}.h^{-1}$.
2. Statistical analysis showed that with an increase in body weight, the aquatic oxygen uptake of *Anabas testudineus* per unit time and per unit body weight respectively increased by 0.4522 and decreased by -0.5726. Both, the increase and decrease in aquatic oxygen uptake to body weight show a highly significant correlation.
3. Similarly, in the early developmental stages of *Channa punctatus*, the aquatic oxygen uptake per unit time gradually increased from 22,482±4.936 to 226.927 ±1.556 $\mu lO_2.h^{-1}$ with a gradual increase in body weight from 8.22±3.89 to 925.50±3.54mg, whereas, the oxygen uptake per unit body weight decreased from 3.107±1.166 to 0.246±0.001 $\mu lO_2.mg^{-1}.h^{-1}$ for the fishes of similar body weight.
4. The relationship between aquatic oxygen uptake and body weight showed that the oxygen uptake per unit time in *Channa punctatus* increased by a

functional power of 0.4064 whereas it decreased by a power of -0.6141 with a unit increase in body weight. Both, the increase and decrease in oxygen uptake has been found statistically significant.

5. Thus, with a unit increase in body weight, the early developmental stages of *Anabas testudineus* consumed more oxygen than that of *Channa punctatus*.
6. Aerial, aquatic and total oxygen uptake during the early developmental stages of *Anabas testudineus* ranged respectively from 4.10±0.16 to 73.30±0.30, 21.20±0.20 to 61.30±0.16 and 25.30±0.36 to 134.60±0.64 $\mu lO_2.h^{-1}$ with an increase in body weight from 56.00±2.00 to 1026.50±6.50 mg. The corresponding weight-specific oxygen uptake was calculated to be 0.0732± 0.009 to 0.0715 ± 0.001 from air, 0.3788 ± 0.0099 to 0.0598 ± 0.002 from water and 0.4520±0.0108 to 0.1313±0.0003 $\mu lO_2.mg^{-1}.h^{-1}$ from both aerial and aquatic sources.
7. The regression analysis showed that with an increase in body weight, the aerial, aquatic and total oxygen uptake of *Anabas testudineus* increased by a functional power of 1.0019, 0.3655 and 0.5801 per unit time, while it showed a decrease by a power of -0.0026, -0.6351 and -0.4200 per unit body weight. Statistically, the increase, as well as decrease in oxygen uptake, was found to be highly significant.
8. Similarly, in the early developmental stages on *Channa punctatus*, with the gradual increase in body weight from 60.00±3.01 to 1040.00±3.50mg, the aerial, aquatic and total oxygen uptake showed an increasing trend respectively from 3.80±0.20 to 45.00±0.30, 24.20±0.30 to 72.60±0.50 and 28.00±0.50 to 117.60±0.80 $\mu lO_2.h^{-1}$. The weight-specific aerial, aquatic and total oxygen uptake for the similar body weights was calculated to be 0.0633 +0.0001 to 0.0433±0.0001, 0.4033±0.0003 to 0.0698±0.0003 and 0.4667±0.0004 to 0.1131±0.0004 $\mu lO_2.mg^{-1}.\ h^{-1}$ showing a decreasing trend.
9. In *Channa punctatus* also, with an increase in body weight the aerial, aquatic and total oxygen uptake increased by a power of 0.9533, 0.3625 and 0.5187 per unit time, while it showed a decrease by functional powers of -0.0466, -0.6368 and -0.4804 per unit body weight. Statistically, the variation in the rate of oxygen uptake to body weight was highly significant.
10. The differences in bimodal oxygen uptake in *Anabas testudineus* and *Channa punctatus* are perhaps related to their different stages of the life cycle, growth pattern, changes in feeding behaviour and other physiological aspects of juveniles and adults.

11. During early developmental stages, *Anabas testudineus* and *Channa punctatus* depend more on the water than air for their total metabolic activities, however, they chiefly depend on air-breathing as they grow in size. This is because with an increase in body weight the diffusing capacity of the gills decreases, leading the larger fishes to resort to aerial breathing. Thus, the efficiency of oxygen uptake of bimodal gas exchange machinery varies mainly with the growth of fishes.
12. Using the relationship, the rate of oxygen uptake by the early developmental stages of *Anabas testudineus* or *Channa punctatus* of particular body weight may be determined. Thus, the actual amount of oxygen required for a given number of fish for transportation may be estimated and applied in practice.
13. With a gradual increase in group size from 5 to 40 fishes, the weight-specific oxygen uptake of *Anabas testudineus* and *Channa punctatus* significantly decreased by a functional power of -0.2647 and -0.2798 respectively under continuously flowing water condition. Thus the oxygen uptake in the early developmental stages of *Anabas testudineus* and *Channa punctatus* was less in grouped conditions than a fish in isolation or a small group.
14. Similarly, under still water conditions, the rate of aquatic oxygen uptake per unit body weight also decreased significantly by a power of -0.7587 in *Anabas testudineus* and by -0.6774 in *Channa punctatus* with an increase in group size.
15. The reduction in oxygen uptake in *Anabas testudineus* and *Channa punctatus* is probably due to a soothing or calming effect since the juveniles of *Anabas testudineus* and *Channa punctatus* prefer to remain in a shoal, sometimes in a group of 3 to 5 juveniles. Their isomorphic juveniles feel easy in a group due to companionship. It is therefore advisable to transport them in a moderate group.
16. The early developmental stages of *Chana punctatus* under the anaesthetic effect of Benzocaine at a concentration of 18.0±3.0 $mg.l^{-1}$ showed a reduction both in opercular frequency and oxygen uptake.
17. Under the controlled condition, a *Channa punctatus* of average body weight 3.71±1.18g consumed 0.097±0.019 $mlO_2.g^{-1}.h^{-1}$ while the 3h and 24h Benzocaine exposed specimen of similar body weight consumed 0.056±0.009 and 0.059± 0.010 $mlO_2.g^{-1}.h^{-1}$ respectively. When compared to controlled conditions, 3h Benzocaine exposed *Channa punctatus* showed 42.43±3.03% reduction in its weight-specific oxygen uptake while 24h exposed fish showed 39.14±2.09% reduction. The reductions are highly significant.

18. The average opercular frequency of 47.0±2.0min^{-1} under controlled conditions for a specimen of *Channa punctatus* goes down to 27.0±2.0 and 29.0±2.0 min^{-1} in 3h and 24h exposed fishes to show 42.81±3.21% and 39.68±2.95% reduction. The reductions were found to be statistically significant.
19. Thus, more early developmental stages of *Channa punctatus* may be transported under anaesthetic effect of a given concentration of Benzocaine.
20. The LC50 dose of ammonia for *Channa punctatus* (body weight: 6.20 to 8.30g) was found to be 1462 mg.l^{-1}. The dose was higher than many fishes probably due to their dual breathing nature.
21. The surface ultrastructure of the ammonia-treated gills showed various alterations in their organizations. Shrinkage, folding and loss of epithelium from gill arch and gill filaments; oozing of blood cells; formation or appearance of some nodule like structures; increase in number and size of mucous glands: deposition of mucus; breakage and loss of micro ridges; the fusion of adjacent lamellae; dissociation and re-association of epithelium and additional outgrowth of filaments were seen under the acidic and alkaline solution of ammonia under short and long term exposure.
22. The increase in mucus secretion forms a coat over the lamellae which adversely affect gaseous exchange due to increased water blood diffusion distance. The micro ridges in the epithelial cells of the gills of *Channa punctatus* probably help in anchorage and spreading of mucus.
23. The fusion of lamellae is probably due to intensified attraction between the epithelium of adjacent lamellae which leads to the formation of an interlamellar bridge. These structures hamper the free movement of lamellae and thus impair proper oxygen uptake.
24. The weight-specific oxygen uptake of a controlled *Channa punctatus* (body weight: 7.176± 0.491g) was calculated to be 0.0192±0.0040 $mlO_2.g^{-1}.h^{-1}$, whereas it was 0.0435±0.0118, 0.0703±0.0090 and 0.0285±0.0086 $mlO_2.g^{-1}. h^{-1}$ in 24h, 48h and 96h of ammonia exposure. Similarly, the opercular frequency of a normal *Channa punctatus* was recorded to be 37.0±3.0 min^{-1} which increased to 44.0 ±3.0, 50.0±4.0 and 43.0±4.0 min^{-1} respectively in 24h, 48, and 96h treatment of ammonia. The changes in oxygen uptake and opercular frequency due to exposure to ammonia were highly significant at 48h of exposure.
25. The increase in oxygen uptake by the ammonia exposed fishes despite the damages to the gills is perhaps due to the additional growth and

bifurcation of filaments which results in increases the respiratory surface area.

26. The increase in opercular frequency is probably related to the increase in oxygen requirement due to increased salt excretion or increased activity of fish due to irritation. During exposure to ammonia, the excessive mucus secretion, fusion of lamellae, extra filamental growth and bifurcation of filaments etc seems to be the defensive responses to combat the adverse effects of the pollutants.
27. A dose of 20.87±6.277 $mg.l^{-1}$ and 22.625±4.269 $mg.l^{-1}$ of Benzocaine was found adequate to suitably anaesthetize *Anabas testudineus* of body weight 7.263±4.364 and *Channa punctatus* of 6.775±4.231g body weight respectively for 96-hour packing.
28. Similarly, an *Anabas testudineus* (average body weight: 10.37±4.73g) requires 60.40±14.10 $mg.l^{-1}$ of MS-222 while 25.875±5.266 $mg.l^{-1}$ of MS-222 is used to sedate *Channa punctatus* mean body weight: 5.488±4.359g.
29. A dose of MS-222 for *Anabas testudineus* and *Channa punctatus* of body weight 14.50±0.5 and 11.55±0.65g was found to be 70.0$mg.l^{-1}$ and 32.50 $mg.l^{-1}$ respectively for long-distance transportation covering 72 hours.
30. It has been calculated that five individuals of *Anabas testudineus* (total weight=72.50g) need 385.880 and 497.977mg oxygen from aquatic and aerial sources respectively for transportation of 72 hours duration. For packing these individuals, 14.113 litre of water (source of aquatic oxygen) and 1.500 litre of commercial oxygen (for aerial oxygen uptake) were filled in the poly bags. 990.0mg MS-222 was dissolved in the water to properly sedate the fish.
31. Similarly, a requirement of 377.182mg oxygen from water and 253.669mg oxygen from the air for five specimens of *Channa punctatus* (body weight = 57.65g) was determined for a 72-hour transportation. For this purpose, a calculated amount of 460.0mg of MS-222, 14.161 litre of water and 1.006litre commercial oxygen was given in the poly bags.
32. In actual practice, some reserve oxygen is provided to cover the delay in transit etc.
33. The process of removing transported fish and placing them in holding tanks at their destination is relatively straightforward, but must be done carefully to minimize fish mortality. Water quality in the transport bags will be dramatically different from that of the holding tank, and the transfer of fish from one to the other should be done slowly to prevent rapid changes in water quality.

10

Flowsheet, Proposals and Future Scope

Flowsheet

Based on various experimentations concerning live fish transportation, a flow sheet may be prepared as:

1. Evaluation of the health status of the fishes purchased from the market to be packed for transportation. Serrated operculum and spiny fins, if any (as in *Anabas testudineus*) is serrated as it damages the packing materials. The trimmed fishes are treated with antibiotics and left for a week for wound healing.
2. In the case of abroad transportation of fish, clearance to the fitness of fishes is made available from a veterinary doctor/inspector. A no-objection certificate is also obtained from the office of the Marine Products Export Organization.
3. Proper sanitization of all the equipment and accessories for the determination of oxygen uptake of fishes as well as all the other steps of fish transportation in this pandemic of COVID-19.
4. Determination of oxygen requirement of the fish in terms of per hour and per unit body weight is essential. In the case of air-breathing fishes to be packed (Like *Channa punctatus*, *Anabas testudineus*, *Clarias batrachus* and *Heteropneustes fossilis*), determination of aerial and aquatic oxygen requirement of the fishes is essential. The oxygen requirement of oxygen was estimated from the calculated values of intercept (=a) and slope (=b). For this calculation, a relationship of aW^b is applied.
5. Evaluation of schooling/shoaling or aggressive/cannibalistic behaviour of the fishes to be packed.
6. Determination of variation in oxygen uptake of fishes to packed in isolation and an association. Percent of decrease or increase in oxygen uptake is to be noted.

7. The optimum dose of an anaesthetic for a particular duration of transport of fish is determined.
8. Determination of variation in oxygen uptake of fishes under anaesthetization. Percent of decrease in oxygen uptake is to be noted.
9. After completion of these parameters, the complete requirement of oxygen of fishes is to be calculated based on total body weight of the fish, behaviour of the fish, variation in oxygen uptake under the association, variation in oxygen uptake under anaesthetization, the suggestion of Saha et al, (1956) for the addition of 25% more oxygen as oxygen reserve and probable en-route delay.
10. The volume of the water (=A) is calculated for a requisite hour of transportation of fishes of known bodyweight following the method of Martyshev (1983). The volume of water (=A) is multiplied by a factor at the desired temperature (=T) to get A x T = B. The value of B is divided by the amount of dissolved oxygen at a particular temperature (Privol'nev, 1964) to find C. The value of C is again multiplied by the weight of fishes (kg) and the duration of transportation (hours) to get an actual volume of water (D).
11. From the value of slope and intercept of aerial oxygen uptake, weight and number of fish consuming 1mg oxygen/hour is calculated. With the help of this value, the required volume of oxygen is filled in the poly bags using an oxygen cylinder.
12. The amount of excretory products (like ammonia in freshwater fishes) of fishes is also calculated and considered to ensure safe packing and transportation.
13. Feeding is stopped before 24 hours of packing, methylene blue treatment is also given to fishes and left for 4 to 5 hours before packing.
14. At the time of packing, sodium chloride is added at the rate of 500mg/l of water.
15. The addition of secondary sodium phosphate or disodium hydrogen phosphate (Na_2HPO_4) in water during packing increases the duration of transport due to its buffering action.
16. Some air space should be left above the water surface in packed containers for safe transport.
17. Fishes once used for transportation may be further used for the same purpose at least after a week.
18. A certificate is obtained from the consigner on reaching the destination including the details of the duration of transport, the health status of fishes and mortality if any.

Proposals

The data in this report will empower administrators to set up and run road transport activities for the live fish market. Proper sanitization of all the equipment and accessories is one of the essential steps for packing and fish transportation in this pandemic of COVID-19. A portion of the characteristic features examined in the report is:

1. Matching of truck size and plan to explicit necessities.
2. Use of protected and insulated tanks.
3. Transporting fish in dim or low light conditions.
4. Design of tanks and stacking frameworks to limit fish dealing with.
5. Provision of oxygen to make up for oxygen devoured by breath.
6. Provision of water agitators to off-gas carbon dioxide.
7. Use of liquid oxygen rather than gaseous oxygen.
8. Provision of satisfactory measures of oxygen during the stacking methodology, when oxygen utilization is most noteworthy.
9. Reduction of water temperature to lessen the metabolic pace of the fish during transport.
10. Extensive pre-transport 'hardening' to adjust the fish to ship conditions.
11. Effects of temperature and fish size on loading rates.
12. The use of plastic bags reduces the weight and chances of injury from hitting the wall of the container.
13. Lindstedt(1914) formula ($R=\frac{V}{N},\frac{V}{N}\alpha W\frac{2}{3}$, where R=rate of respiration, V= oxygen entering h^{-1} into a volume of water, N=number of fry or fingerlings and W= weight of fry) may be applied for the transport of a consignment of fry or fingerlings of fishes.

Future Scope

Studies on live fish transportation have suggested the importance of considering variations in osmoregulatory and water-quality sensitivity among various fishes namely marine as well as freshwater species, during the designing of live transportation protocol. Proper sanitization of all the equipment and accessories is one of the essential steps for the determination of oxygen uptake of fishes as well as all the other steps of fish transportation in this pandemic of COVID-19. The individual diverseness within species is also responsible and depends on various factors like health status, life stage etc. Hence the protocols prior to, during as well as after transportation play an important role in addressing the success of live fish transports.

Extensive studies need to be carried out from this perspective for efficient transportation protocols. Furthermore, in India, a customized design of live fish containers to facilitate the transportation of a variety of fishes under economic conditions is currently not available.

Hence research and development need to be focused on this aspect to bring about a compact device considering the economic factors to meet the demand of seafood consumers in different sectors of society.

Filling this void can facilitate the supply of fresh and quality fish to the customers in various parts of the Country. Intense research is being carried out by ICAR-CIFT in this regard for the design and development/construction of a model live fish transportation system.

It will benefit to Fish traders, Fish raisers, Fish farmers as well as Aqua-culturists by this latest technology of the live fish transportation. Therefore, it is expected that the technology will help to raise the GDP of the country as a whole.

References

Abrahams, M.V. and Colgan, P. (1985): Risk of predation, hydrodynamic efficiency and their influence on school structure. Environ. Biol. Fishes 13: 195–202.

Adhikari, S., Sinha, A.K., Roy, P.K. and Munshi, J.S.D. (1997): Ultrastructural changes in the gill – lamellae of a catfish, *Heteropneustes fossilis* after sublethal exposure to different concentrations of artificial sea-water. J. Freshwater Biol. 9(1): 29 – 35.

Agrawal, S. K., Rai, A. K., Banerhee, T. K. and Mittal, A.K. (1979): Histopathology of the epidermal mucous cell in relation to salinity in a freshwater teleost, *Heteropneustes fossilis* (Blooh) (Heteropneustidae, Pisces). Zool. Beitr. 25: 403 – 410.

Alabaster, J.S. and Herbert, D.W.M. (1954): Influence of carbon dioxide on toxicity of ammonia. Nature. 174: 404 – 405

Alikunhi, K. (1957): Fish culture in India. Fm. Bull. Indian Coun. Agri. Res., 20: 144.

Allee, W.C. (1931) *Animal aggregation a study in General Sociology.* University of Chicago Press, Chicago.

Anderson, R. (1979): The effect of ammonia on the growth of juvenile dover sole (*Solea solea*) and turbot (*Scophtalmus mexicum*). Aquacult., 17: 291 – 309

Anees, M.A. (1975): Acute toxicity of four organophosphorous insecticides to a freshwater teleost *Channa punctatus* (Bloch). Pakistan J. Zool., 7: 135 – 141.

Anon. (2020): Corona virus found in Paris industrial water. *Corona24 News*. Digital. https://www.corona24.news/c/2020/04/19/coronavirus-found-in-parisindustrial-water.html, April 19, 2020.

Aravindakshan, P.K., Singh, S.K., Sarkar, B., Majhi, D., Jena, J.K., Pani, K.C., Muduli, H.K., Ayyappan, S., B. C. Mohapatra, B.C. and Sarangi N. (2011): Oxygen consumption of carps during live transportation. Indian J. Fish., 58(4): 67-70.

Aronson, L. R. (1963): The central nervous system of sharks and bony fishes with special reference to sensory and integrative mechanism. In *Sharks and Survival* (Ed. P. W. Gilbert). pp. 165-241. Heath, Boston, Massachusetts.

Barbieri, E. and Bondioli, A.C.V. (1993): Estimation of acute toxicity of ammonium sulphate of freshwater catfish, *Heteropneustes fossilis*. II. A. Histopathological analysis of the epidermis. Biomed. Environ. Sci., 6: 45 – 58.

Barbieri E. and Bondioli A.C.V. (2015): Acute toxicity of ammonia in Pacu fish (*Piaractus mesopotamicus*, Holmberg, 1887) at different temperatures levels. Aquaculture Research, 46 (3): 565-571.

Barlow, G.W. (1961): Intra and inerspecific differences in rate of oxygen consumption in gobidd fishes of the genus *Gillichthys*. Bio. Bull., 121: 209 – 229

Barrionuevo, W.R. and Fernandes, M.N. (1908): Time course of respiratory metabolic adjustments of a South American fish, *Prochilodus scrofa*, exposed to low and high temperatures. J. App. Ichthyol., 14(1-2): 37 – 41.

Barton, B. A. (1997): Stress in finfish – a historical perspective. In: Fish Stress and Health in Aquaculture (Iwama, G. W., Pickering, A. D., Sumpter, J. P. and Schreck, C. B., Eds), pp 1–33, Cambridge University Press, New York

Basavaraju, Y. Mair, G.C. Mukthayakka, G. Devi, B.S.R. and Penmen, D.J. (1998): The use of benzocaine as an anaesthetic and sedative for carps. Proc. Nat. Acad. Sci. India., 68 (B): III and IV: 231 – 240.

Basu, S.P. (1950): Some experimental data regarding the oxygen requirements of the Indian fishes, *Catla catla, Labeo rohita, Labeo bata* and *Cirrhina mrigala*. Proc. Nat. Inst. Sci. India, 15B (1): 283 – 289.

Basu, S.P. (1951): Effect of addition of soil in transport on larvae of Indian carps. Proc. Indian Sci. Congr., 38(3): 227.

Baumert, M. (1853): Chemische Untersuchungen über die Respiration des Schlammpeizgers (*Cobitis fossilis*). Annalen der Chemie und Pharmacie 88, 1–56.

Beamish, F.W.H. (1964a): Respiration of fishes with special emphasis on standard oxygen consumption. II. Influence of weight and temperature on respiration of several species. Can. J. Zool., 42: 177 – 188.

Beamish, F.W.H. (1964b): Seasonal changes in the standard rate of oxygen consumption of fishes. Can. J. Zool. 42: 189 – 194.

Belding (1929): The respiratory movements of fish as an indicator of a toxic environment. Trans. Amer. Fish. Soc., 59: 238 – 245.

Bell, G.R. (1967): A guide to the properties, characteristics and uses of some general anaesthetics for fish. Bull. Fisheries Res. Board Can., 148: 1 – 4.

Belokopitin, I.S. (1968): Vopr. Intiol., 8(2): 382 – 385.

Bennett, M.B. (1988): Morphometric analysis of the gills of the European eel, *Anguilla anguilla*. J. Zool., Lond., 215: 549 – 560.

Bhakta J.N. (2006): Ammonia Toxicity to Four Freshwater Fish Species: *Catla catla, Labeo bata, Cyprinus carpio* and *Oreochromis mossambica*. Electronic Journal of Biology. 2(3): 39-41.

Bhasker, M. and Govindappa, S. (1985): Physiological and metabolic adaptation of fish, *Tilapia mossambica* (Peters) on acclimation to altered pH media. Ambio., 14(6): 349 – 351.

Bhaumik, U., Pandit, P.K. and Banerjee, B.K. (1986): Finfish and Shellfish transport. In: Second National Fish Seed Congress, July 3-4, 1986, Calcutta, p. 99-101.

Bhoopathy, S. and Gunesegar, N. (1999): Potassium dichromate induced physiological changes at sub – lethal concentration in the exotic fish *Oreochromis mossambicus* (Trewavas). Proc. Nat. Acad. Sci. India, 69 (B): 251 – 257.

Bin, X., Kaijin, W. and Xiangyun, Z. (2018): Effects of MS-222 and eugenol on oxygen consumption rate, ammonia excretion rate and asphyxiation point of juvenile *Procambarus clarkii*, Journal of Dalian Ocean University, 33(4): 455-459.

Biswas, N., Ojha, J. and Munshi, J.S.D. (1979): Bimodal oxygen uptake in relation to body weight of the amphibious mud – skipper, *Boleophthalmus boddaerti* (Pall.). Indian J. Exp. Biol., 17: 722 – 756.

Black, E.C. (1958): Hyperactivity as a lethal factor in fish J. Fish. Res. Bd. Can., 15: 573 – 586.

Bloch, M. E. (1785 – 1795): Naturegeschichte der auslandischen fish, Berlin, 9 parts in 2, and atlas (3 volumes).

Bodgan, E. (1972): Przewożenie narybki letniego sielawy, siei i pelugi. (Transport of the yearlings of great maraena, vendace, peled). Publ. Inst. Rybaetwa Sródladowego, Olsztyn, (55):16 p.

Bove, J. (1962): MS – 222 Sandoz: The anaesthetic of choice for cold blooded organisms. Sandoz News, No. 3. 12p.

Bowen, E.S. (1932): Further studies of the aggregation behaviour of *Ameriurus melas.* Biol. Bull., 17: 258 – 270.

Brett, J.R. (1965): The relation of size to rate of oxygen consumption and sustained swimming speed of sockeye salmon (*Oncorhinus merka*). J. Fish. Res. Bd. Can., 22: 1491 – 1501.

Brett. J.R. and Sutherland, D.B. (1965): Respiratory metabolism of pumpkinseed (*Lepomis Gibbosus*) in relation to swimming speed. J. Fish. Res. Bd. Can., 22: 405 – 409.

Brockway, D.R. (1950): Metabolic products and their effects. Prog. Fish. Cult., 12: 127 – 129.

Brown, L. (1987): *Lab. Anim.* 21: 210.

Buckley, J.A. (1978): Acute toxicity of un-ionized ammonia to fingerling coho salmon.: Prog. Fish. Cult. 40: 30 – 32.

Bullock, G.L. (1972): Studies on selected myxobacteria pathogenic for fishes and on bacterial gill disease in hatchery reared salmonids. U.S. Fish Wildl. Servo Tech. Pap. No. 60: 30 p

Burggren, W.W. (1979): Bimodal gas exchange during variation in environmental oxygen and carbon dioxide in the air breathing fish *Trichogaster trichopterus.* J. Exp. Biol. 82:197 -213.

Burkhalter, D.E. and Kaya, C.M. (1977): Effect of prolonged exposure to ammonia on fertilized eggs and sac fry of rainbow trout. Tran. Am. Fish. Soc. 106: 470-475.

Burrows, R.E. (1949): Recommended methods for fertilization, transportation and care of salmon eggs. Prog. Fish. Cult., 11: 175 – 177.

Burrows, R.E. (1964): Effects of accumulated excretory products on hatchery-reared salmonids. U. S. Bur. Spt. Fish. Wildl. Res. 66: 1 – 11.

Bushell, P.G., Steffensen, J.F., Schurmann, H. and Jones, D.R. (1994): Exercise metabolism in two species of cod in arctic waters. Polar Biol., 14: 43 – 48.

Cairns, J., Heath, G.A. and Parker, B.C. (1975): The effects of temperature upon the toxicity of chemicals to aquatic organisms. Hydrobiol. 47: 135 – 137.

Calamari, D., Marchetti, R. and Vailati, G. (1981): Effects of long term exposure to ammonia on the developmental stages of rainbow trout (*Salmo gairdneri*). In "*The Early Life History of Fish*" (Eds. Lasker, R. and Sherman, K.) pp. 81–86. Conscil Int. Explor. Mer. Copenhagen,

Cameron, J.N. and Heisler, N. (1983): Studies of ammonia in the rainbow trout-physicochemical parameters, acid-base behaviour and respiratory clearance. J. Exp. Biol. 105: 107–125.

Cardoso, E. L., Chairini-Garcia, H., Ferreira, R.M.A. and Poli, C.R. (1996): Morphological changes in the gills of *Lophiosilurus alexandri* exposed to un-ionized ammonia. J. Fish Biol., 49: 778 – 878.

Carmichael, G. J., Tomasso, J. R. and Schwelder, T. E. (2001) Fish transportation. In: Fish Hatchery Management (Wedemeyer, G.A., Ed), pp 641–660, American Fisheries Society, Bethesda, MD

Cech, J.J.Jr. (1990). Respirometry. In Methods for Fish Biology (Schreck, C. B. & Moyle, P. B., eds), pp. 335–362. Bethesda, MD: American Fisheries Society.

Chakrabarty, R.D. and Murty, D.S. (1972): Life history of Indian major carps, *Cirrhinus mrigala* (Ham.), *Catla catla* (Ham.) and *Labeo rohita* (Ham.) J. Indian Fish Soc. India., 4: 132–161

Chan, P.K.Y. (1990): Aerial and aquatic oxygen consumption rates of Hong Kong mudskippers. Asian Marine Biology., 7: 180 – 199.

Chanchal, A.K. and Pandey, B.N. (1979): Effect of gonadal steroids on respiratory metabolism of *Anabas testudineus* (Bloch). Endokrinologie., 74(3): 297 – 303.

Changlin, L., Li, H. and Siqing, C. (2007): A Review of Fish Anesthesia Research .Fisheries Modernization, 34(5):21-25.

Chaohui, Z., Jiaori, C. and Wang B. (2003): The effect of anesthetic eugenol on oxygen consumption of *Trachinotus blochii* .Oceanography, (6):11-14.

Chen, X-J. (2019): Research on the Application of Fishery Anesthetic MS-222 and Eugenol in China. International Journal of Innovative Studies in Aquatic Biology and Fisheries. 5(4): 16-20.

Choudhary, D.P. (1979): *Some aspects of respiratory physiology of Channa striata (Bloch) (Channaidae: Channiformes)*. Ph. D. Thesis, Bhagalpur University, Bhagalpur, India.

Christie, R.M., and Battle, H.J. (1963): Histological effects of 3-trifluoromethyl-4- nitrophenol. (TFM) on larval lamprey and trout. Can. J. Zool. 41:51-61.

Chun, D., Lei, P. and Ye, Q. (2016): Overview of fishery anesthetics .Aquatic scientific and technological information. 43(01):28-32+36.

Colesante, R.J. (1977): Improvements of osocid culture techniques. New York Department of Environmental Conservation, Report of 1976 studies, Albany.

Collymore, C., Tolwani, A., Lieggi, C. and Rasmussen, S. (2014): Efficacy and safety of 5 anesthetics in adult zebrafish (*Danio rerio*). J Am Assoc Lab Anim Sci. 53(2):198–203.

Colt, J. and Tchobanoglous, G. (1976): Chronic enposure of channel catfish, *Ictalurus punctatus*, to ammonia: Effects on growth and survival. Aquaculture. 15: 353–372.

Crawford, R.E. (1978): Different oxygen consumption rates by the longhorn sculpin *Myoxocephalus octodecimspinosus* in an apparatus designed for shipboard use. Mar. Biol., 44(4): 377 – 381.

Cruz–Neto, A.P. and Steefensen, J.F. (1997): The effects of acute hypoxia and hypercapnia on oxygen consumption of the freshwater European eel. J. Fish Bio., 50: 759 – 769.

Danecker, E. (1964): Die Jauchevergiftung von Fischen-eine Ammoniakvergiftung. Osterreichs Fischerei., 3: 55 – 68.

Das, P.C., Ayyappan, S., Jena, J.K. and Das, B.K. (2004): Acute toxicity of ammonia and its sub-lethal effects on selected haematological and enzymatic parameters of mrigal, *Cirrhinus mrigala* (Hamilton). Aquaculture Research, 35(2):134.

Das, P. (1990): AFSIB. News Letter. IV (I): 6 – 8.

Das, P. and Bhaumik, U. (1979): Transport of live fish for stocking. Souvenir: Commemoration of Golden jubilee of ICAR, CIFRI, Barrackpore, Part-1, p. 64-68.

Daye, P.G. and Garside, E.T. (1976): Histopathological changes in superficial tissues of brook trout. *Salvelinus fontinalis* (Mitchell) exposed to acute and chronic levels of pH. Can. J. zool. 54: 2140-2155.

De Graaf, P.J.F., Ballintijn, C.M. and Maes, F.W. (1987): Machanoreceptor activity in the gills of the carp. I. Gill filament and gill raker mechanoreceptor. Respir. Physiol., 69: 173 – 182.

Dejours, P. (1975): *Principles of comparative respiratory Physiology*. North Hill and American Elsevier, 252p.

Delco, E.A. (1963): Reduced metabolic rates in males of two cyprinid fishes. Copeia, 176 – 177.

Divelly, J.H., Mudge, W.N. and Antony, A. (1977): Blood pO_2 and pH changes in brook trout, *Salvelinus fontinalis* exposed to sublethal levels of acidity Comp. Biochem. Physiol., 57A: 347 – 351.

Domenici, P., Ferrari, R.S., Steffensen, J.F. and Batty, R.S. (2002): The effect of progressive hypoxia on school structure and dynamics in Atlantic herring *Clupea harengus*. Proc. R. Soc. Lond. B 269: 2103–2111.

Dommasnes, A., Rey, F. and Rottingen, I. (1994): Reduced oxygen concentration in herring wintering areas. ICES J. Mar. Sci. 51: 63–69.

Dong, X., Zhang, X., Qin, J. and Zong, S. (2013): Acute ammonia toxicity and gill morphological changes of Japanese flounder *Paralichthys olivaceus* in normal versus supersaturated oxygen. Aqua. Res. 44(11): 1752-1759. https://doi.org/10.1111/j.1365-2109.2012.03181.x.

Dupree, H.K. and Huner, J.V. (1984): The status of warmwater fish farming and progress in fish farming research. Washington, D.C., U.S. Fish and Wildlife Service, pp. 165-176.

Dutta, H.M., Zeno, T. and Richmonds, C. (1992): Toxic action of diazinon on the gills of blue gill sunfish, *Lepomis macrochirus*. Ohio. J. Sci., 92(2): 9.

Eddy. F.D. (1982): Osmotic and ionic regulation in captive fish with particular reference to salmonids. Comp. Biochem. Physiol., 72B: 125 – 141.

Edwards, R.R.C., Blaxter, J.H.S., Gopalan, U.K., Mathew, C.V. and Finlayson, D.M. (1971): Feeding metabolism and growth of tropical flat fish J. Exp. Mar. Biol. Ecol., 6(3): 279 – 300.

Emerson, K., Russo, R. C., Lund, R. E. and Thurston, R. (1975): Aqueous ammonia equilibrium calculations: Effects of pH and temperature. J. Fish. Res. Bd. Can., 32: 2379 – 2382.

European Inland Fisheries Advisory Commission. (1970): Water quality criteria for European freshwater fish. Report on ammonia and inland fisheries. EIFAC Tech. Paper. Rome: FAO., 11: 12 p.

Evans, D. H. and More, K. J. (1988): Modes of ammonia transport across the gill epithelium of the dogfishpup (*Squalus acanthias*). J. exp. Biol. 138: 375 – 397.

Evans, D. and Cameron, J. (1986): Gill ammonia transport. J. Exp. Zool., 239: 17 – 23.

Farmer, G. J., and Beamish F. W. H.(1969): Oxygen consumption of *Tilapia nilotica* in relation to swimming speed and salinity. J. Fish. Res. Bd Can. *26*, 2807 – 2821.

Fereira, J.T., Schoonbee, J.J. and Smit, G.L. (1984): The use of benzocaine-hydrochloride as an aid in the transport of fish. Aquaculture, 42(2):169-74

Finney, D. J. (1964): *Probit analysis*. 2nd ed. Cambridge University Press, U. K. 20p.

Flis, J. (1968a): Anatomicohistopathological changes induced in carp (*Cyprinus carpio* L.) by ammonia water. I. Effects of toxic concentrations. Acta. Hydroboil. Sin., 10: 205 – 224.

Flis, J. (1968b): Anatomicohistopathological changes induced in carp (*Cyprinus carpio* L.) by ammonia water. II. Effects of subtoxic concentrations. Acta Hydrobiol. 10: 225 – 238.

Fromm, P. O., and Gillette, J. R. (1968): Effect of ambient ammonia on blood ammonia and nitrogen excretion of rainbow trout (*Salmo gairdneri*). Comp. Biochem. Physiol., 12: 489 – 499.

Fry F. E. J., Hart J. S., (1948): The relation of temperature to oxygen consumption in the goldfish. Biol. Bull., 94: 66 – 77.

Fry, E. F. J. (1947): Effect of the environment on animal activity. Univ. Toranto Studies. Biol. Ser. 55. Publ. Ontario Fisheries Res. Lab., 68: 1 – 62.

Fry, F. E. J. (1971): The effect of environmental factors on the physiology of fish. In: *Fish physiology* (Eds. W. S. Hoar and D. J. Randall), pp. 1- 98. Vol. VI, Academic Press, New York.

Fulin, Z. (2008): Study on anesthetic effect of eugenol on carp [D]. Guizhou University.

Ganapati, S. V. and Chacko, P. I. (1951): A comparative study of the transport of fish seed in oxygen tin carrier and in ordinary tin carrier. Ind. Comm. J. 6: 564 – 587.

Gangqiao, K., Ping, L. and Shuming, Z. (2010): Anesthetic effect of eugenol on young *Ictalurus punctatus*. Sichuan Journal of Zoology. 29(5): 584-587.

Geyer, F. and Mann, H. (1939): Beitrage zur Atmung der Fische. III. Der Sauerstoffverbrauch in Gruppenversuch. Z. Vergl. Physio., 27: 429 – 433.

Ghosh, T. K. (1984): *Bimodal gas exchange of some air – breathing teleostean fishes*. Ph. D. Thesis, Bhagalpur University, Bhagalpur. 131p.

Ghosh, T. K. and Biswas, N. (1980): Rhythmic behaviour in the bimodal oxygen uptake of *Anabas testudineus* (Bloch). Biol. Bull. India., 2(1): 8-11.

Ghosh, T. K. and Jha, K. K. (1994): Packing and long distance transportation of certain live air – breathing teleosts from India. Abs. 124 Ann. Met. Am. Fisheries Soc. 21 – 25. Aug. Halifax (Nova Scotia), Canada. p – 91.

Ghosh, T. K., Biswas, N. and Munshi, J. S. D. (1980): Oxygen uptake in relation to body weight in swamp eel *Monopterus (=Amphipnous) cuchia* (Ham.), (Symbranchiformes, Amphipnoidae). Ind. Biol. Jour. 2(2): 49 – 54.

Ghosh, T. K., Jha, K. K. and Pandit, D. N. (1994): Effect of group behaviour on the aquatic oxygen uptake in the juveniles of siluroid fish *Heteropneustes fossilis* (Bloch). Abs. Congress Societas Europea Ichthyologroum. Oviedo, Spain, Symposium C 0 : 22.

Ghosh, T. K., Kunwar, G. K. and Munshi, J. S. D. (1990): Diurnal variation in the bimodal oxygen uptake in air – breathing catfish, *Clarias batrachus*, Japan. J. Ichthyol., 37(1): 56 – 59.

Ghosh, T.K., Moitra A. and Munshi J. S. D. (1986): Bimodal oxygen uptake in a freshwater air-breathing fish, *Notopterus chitala*. Japan. J. Ichthyol., 33 (3): 280-285.

Ghosh, T.K. and Moitra A. and Munshi J. S. D. (1990): Bimodal oxygen uptake in relation to body weight and seasonal temperature of on air-breathing murrel, *Channa punctatus* (Bloch). J. Freshwater Biol., 2 (1): 57-64.

Ghosh, T.K. and Munshi, J. S. D. (1987): Bimodal oxygen uptake in relation to body weight and seasonal temperature of an air-breathing climbing perch, *Anabas testudineus* (Bloch). Zool. Beitr., 31 (3): 357-364.

Goel, K. A. and Sastri, K. V. (1995): *A Text Book of Animal Physiology*. Rastogi Publications, Meerut. pp. 234 – 239.

Goldstein, L., Claiborne, J. B. and Evans, D. E. (1982): Ammonia excretion by the gills of two marine teleost fish: the importance of permeance. J. Exp. Zool., 219: 395 – 397.

Gomes, L.C., Araujo-Lima, C.A.R.M., Chippari-Gomes, A.R. and Roubach, R. (2006): Transportation of juvenile tambaqui (*Colossoma macropomum*) in a closed system. Braz J Biol. 66(2A): 493–502.

Gopalakrishnan, V. (1968): Diseases and parasites of fishes in warm water ponds of Asia and the Far East. FAO Fish Rep., 44(5): 319 – 344.

Gordan, M. S., Boetius, I. Evans, D. H., Mc Carthy, R. and Oglesby, L. C. (1969): Aspects of Physiology of terrestrial life in amphibious fishes. I. The mudskipper, *Periothalamus Sobrinus*. J. Exp. Biol., 50: 141 – 149.

Gordon, M.S., Ng, W. W. S. and Yip, A.Y. - W. (1978): Aspects of the physiology of terrestrial life in amphibious fishes.-III. The Chinese mudskipper, *Periophthalmus cantonensis*. J. Exp. Biol., 72: 57-75.

Graham, J. B. (1997): Air-Breathing Fishes. San Diego, CA: Academic Press.

Grigor'eva, M. B. (1966): Effects of Shoaling on respiratory metabolism of fishes, Report summary of the All Union Conference on the Ecology and Physiology of Fishes, Moscow: 130 – 131.

Guangpeng, F., Zhuangping. and Longzhen, Z. (2010): Effects of anesthetic eugenol on oxygen consumption rate of juvenile *Acipenser baeri* and blood biochemical parameters. Journal of Dalian Fisheries University. 25(2): 113-118.

Guo, F.C., Teo, LH. and Chen, T.W. (1995): Effects of anaesthetics on the oxygen consumption rates of platyfish, *Xiphophorus maculatus* (Günther). Aqua Res. 26(12): 887-894.

Hakim, A., Munshi, J. S. D. and Hughes, G. M. (1978): Morphotmetrics of the respiratory organs of the Indian green snake–headed fish, *Channa punctatus*., J. Zool. Lond., 184: 519 – 543.

Hakim, A., Munshi, J. S. D. and Hughes, G. M. (1983): Oxygen uptake from water through the respiratory organs in *Channa punctatus* (Bloch) in relation to body weight. Proc. Indian Natn. Sci. Acad., B 49 (2): 73 – 85.

Hamilton, B. (1822): An Account of the Fishes found in the River Ganges and its Branches. 405p. Archibad Constable and Company, Edinburgh.

Harmon, T. S. (2009) Methods for reducing stressors and maintaining water quality associated with live fish transport in tanks: a review of the basics. Rev. Aquacult. 1(1): 58-66

Helfman, G., Collette, B. and Facey, D. (1997): The Diversity of Fishes, Blackwell Publishing. p 375.

Hemingsen, A. (1960): Energy metabolism as related to body size and respiratory surfaces and its evolution. Rept. Steno. Mem. Hosp. Nord. Insulin Lab., 9: 1 – 110.

Hill, J.V. and Forster, M.E. (2004): The effects of settlement and anaesthesia on oxygen consumption in the New Zealand spotty (*Notolabrus celidotus*) measured in an automated flow-through respirometry system. New Zealand Natural Sciences. 29: 39-48

Hoar, W. S. (1966): *General and Comparative Physiology*. Prentice – Hall Inc., Englewood cliffs, New Jersey, 815 p.

Hora, S. L. (1951): Knowledge of the ancient Hindus concerning fish and fisheries of India. Matsyavinoda or a chapter on angling in the Manosoltara by king Somesvara (1127 A.D). J. Asiat. Soc., 17(2): 145 – 169.

Hora, S.L. (1956): Fish paintings of third millennium B.C. from Nal (Baluchistan) and their Zoogeographic significance. Mem, Indian Mus., 14(2): 78-84.

Hora, S.L. and Pillay, T. V. R. (1962): *Hand-book on fish culture in the Indo-Pacific region*. FAO Fish. Biol. Tech. Pap., 14: 204 p.

Houlihan, D.F. and Morrison, D.J. (1982): Aerial and aquatic respiration in some littoral fish in relation to their vertical zonation. Gas Exchange, Gas Transport and Acid base Regulation in lower vertebrates. Symposium Max-Plank-Institut. Fur experimentelle Medizin, Gottingen (abstract).

Houston, A.H., Madden, J.A., Woods, R.J. and Miles, H.M. (1971): Some physiological effects of handling and tricaine methanesulphonate anesthetization upon the brook trout (*Salvelinus fontinalis)*. J. Fish. Res. Bd. Can. 28: 625–633.

Howarnd, R.M. and Schoettger, R.A. (1969): Efficacy of methylpentynol as an anaesthetic on four salmonids. Invest. Fish Control.No. 29. U. S. Sport Fish. Wildl., pp -1-15.

Hughes, G.M. (1972): Morphometrics of fish gills. Respir. Physiol. 14: 1-25.

Hughes, G.M. (1979): Scanning electron microscopy of the respiratory surfaces of trout gills. J. Zool., 188:, 443–453.

Hughes, G. M. and Singh, B. N. (1971): Gas exchange with air and water in an air – breathing catfish, S*accobranchus (=Heteropneustes) fossilis*. J. Exp. Biol., 55: 667-682.

Hughes, G.M. and Singh, B.N. (1970a): Respiration in an air-breathing fish, The climbing perch *Anabas testudineus*. I. Oxygen uptake and carbon dioxide release into air and water. *J. Exp. Biol.* 53: 265 – 280.

Hughes, G.M. and Singh, B.N. (1970b): Respiration in an air-breathing fish, The climbing perch *Anabas testudineus*. II. Respiratory patterns and their control of breathing. J. Exp. Biol. 53: 281 -298.

Hughes, G.M., Munshi, J.S.D. and Ojha, J. (1986): Post – embryonic development of water and air – breathing organs of *Anabas testudineus* (Pisces: Anabantidae). J. Zool. Lond., 170: 201-225.

Hughes, G.M. and Perry, S.F. (1976): Morphometric study of trout gills: Ä light microscopic method suitable for the evaluating of pollutant action. J. Exp. Biol. 64: 447—460.

Hughes, G.M. and Wright, D. E. (1970): A comparative study of the ultrastructure of the water/ blood pathway in the secondary lamellae of teleost and elasmobranch fishes, Benthic forms. Z. Zellforsch., 104:478-493.

Hughes, G.M., Dube, S.C. and Munshi, J.S.D. (1973): Surface area of the respiratory organs of the climbing perch, *Anabas testudineus*. J. Zool. Lond., 170: 227-243.

Huish, M.T. (1972): Some responses of the brown bullhead to MS_{222}. Proc. Fish-Cult.,34(1):27-32.

Imabayashi, H. and Takahashi, M. (1987): Oxygen Consumption of Post-larval and Juvenile Red Sea Bream, *Pagrus major*, with special Reference to Group Effect. J. Fac. Appl. Biol. Sci., 26: 15-21.

Indian Council of Agriculture Research (1994): FTR of Physiological studies in relation to transportation of certain carp and air–breathing fishes. 36p.

Ioshev, L. (1980): Transportirane na zhiva riba u nas - problemi i normi (Live fish transport in our fish culture - problems and standards). Izv. Inst. Sladkovodno Ribovod., Plovdiv, 14:35–42

Itazawa, Y., Matsumoto, T. and Kanda, T. (1978): Group effects on physiological and ecological phenomena in fish, I. Group effect on the oxygen consumption of the rainbow trout and Medaka. Bull. Jap. Soc. Sci. Fish. 44(9): 965–969.

Jacobs, E., Esmond, E. Melisky, E. and Eocutt, C. (1981): Morphological changes in gill epithelia of heat stressed rainbow trout *Salmo gaindneri*: Evidence in support of a temperature induced surface area change hypothesis. Can. J. Fish Agril. Sci., 38: 16.

Jagannadhan, N. (1947): A note on the collection, conditioning and transport of fingerlings of catla in Madras Presidency. J. Bombay Nat. Hist. Soc., 47(2): 315- 319.

Jager, S.D. and Dekkers, W.J. (1975): Relationships between gill structure and activity in fish. Neth. J. Zool., 25(3): 276-308.

Jha, K.K., Pandit, D.N. and Ghosh, T.K. (2004): A comparative study on aquatic oxygen uptake under sedation in early developmental stages of *Catla catla* (Ham) and *Labeo rohita* (Ham). J Indian Fish Soc India. 36(2): 61-64.

Jha, K. K. (1995): *Physiological studies in relation to transportation of certain Indian major carps*. Ph. D. Thesis, Bhagalpur University, Bhagalpur. India. 110p.

Jha, K. K. and Ghosh, T.K. (1993): Oxygen uptake in the juveniles of an Indian major carp, *Labeo rohita* (Ham.) under an anaesthetic, Benzocaine. Proc. Nat. Acad. Sc. India Bangalore. P-16.

Jha, K.K. and Ghosh, T. K. (1991): Effect of salinity on the oxygen uptake of the fingerlings of *Labeo rohita* (Ham.). J. Freshwater Biol., 3(4): 315 – 318.

Jhingran, V.G. (1975): *Fish and fisheries of India*. Hindustan Publishing Corporation (India), Delhi. 954 pp.

Jhingran, V.G. (1985): *Fish and fisheries of India*. Hindustan Publishing Corporation (India), Delhi. 727 pp.

Jhingran, V.G. (1991): *Fish and fisheries of India*. Hindustan Publishing Corporation (India), Delhi. 727 pp.

Jhingran, V.G. and Pullin, R.S.V. (1985; revised 1988): A Hatchery Manual for the Common Chinese and Indian Major Carps. Asian Development Bank and International Centre for Living Aquatic Resources Management. Manila., p 191.

Jiandong, Z., Gang, C. and Baogui, T. (2010): Effects of anaesthetic MS222 on oxygen consumption and ammonia excretion rate of juvenile Nile tilapia. Fisheries modernization, 37(4):35-39.

Jiayuan, T., Jingshu, W. and Jianyim W. (2011): Effects of different treatments of MS-222 on anesthetic effects of *Acipenser schrenckii* and *Acipenser sinensis*. Journal of Aquatic Ecology, 32(5):87-90.

Jie, B., Hongbo, J. and Huanyun, L. (2014): Study on the anesthetic effect of clove oil on Loach. 45(05): 621-624.

Job, S.V. (1955): The oxygen consumption of *Salvelinus fontinalis*. Univ. Toronto Stud. Biol., 61:73: 1-39.

Johansen, K. and Lenfant, C. (1967): Respiratory function in the South American lung fish, *Lepidosiren paradoxa* (Fitz). J. Exp. Biol. 46: 205-18.

Jones, J.D. (1972): Chapter-5. In *Comparative physiology of Respiration*. pp. 22-33. Edwards Arnold, London.

Jones, J.R.E. (1964): Fish and river pollution. Butter worth, London.

Jones, M.P., Holliday, F. G. T. and Dunn, A.E.G. (1996): The ultrastructure of the epidermis of larvae of the herring (*Clupea harengus*) in relation to the rearing salinity. J. Mar. Biol. Assn., U. K., 46: 235-239.

Joy, K. P. and Sathyanesan, A.G. (1977): Ammonium sulphate as a thyroid inhibitor in the freshwater teleost *Clarias batraclus* (L). Curr. Sci., 46: 671-672.

Jozuka, K. (1966): Chloride excreting and mucous secreting cells in the gills of Japanese Common Eel, *Anguilla japonica*. Annot. Zool. Japan., 39: 202 – 210.

Kamler, E. (1972): Respiration in carps in relation to body size and temperature. Pol. Arch. Hydrobiol., 19: 325-331.

Kamler, E. (1976): Variability of respiration and body composition during early developmental stages of carp. Pol. Arch. Hydrobiol., 23(3): 431 – 485.

Kanda, T. and Itazawa, Y. (1981): Group effect on oxygen consumption and growth of the catfish eat. Bull. Japan. Soc. Sci. Fish., 47(3): 341-345.

Kanda, T. and Itazawa, Y. (1986): J. Fac. Agr., Kyushu Univ., 30(4): 197 – 207.

Kannedy, G.J.A. and Pitcher, T.J. (1975): Experiments on Homing in Shoals of the European Minnow, *Phoxinus Phoxinus* (L.), Trans. Am. Fish. Soc., 104: 452-455.

Kewalramani, H.G. and Gogate, M.G. (1968): Anaesthetization in fish, Tilapia and major carps. Proc. Indian Acad. Sci., 67B(5): 237 – 246.

Khameleva, K.N. (1973): Energy metabolism of *Idotea baltioa bastin* (Aud). Tr. Bses. Gidrobiol. Olesch., 18: 527.

Khan, H. (1939): Seasonal Notes. Punjab Agricultural College, Lyallpur. 18A (2): 49-50.

Khan, H. (1940): Fish ladders in the Punjab. J. Bombay nat. Hist. Soc., 41(3):551 – 562.

Khan, H., (1946): Transport of live-fish in oxygenated containers. Curr.Sci., 15:51

Khanna, S. S. (1998): *An introduction to Fishes*. Central Book Dept. Allahabad, p.171.

Kidd, R.B. and Banks, G.D. (1990): Anaesthetizing lake trout with Tricane (MS_{222}) administered from a spray bottle. The Prog. Fish-Cult., 52: 272 – 273.

Kinsella, E. (1986): Food components with potential therapeutic benefits. The n-polyun saturated fatty acids of fish oils. Food Technology. 40(2): 88-97.

Klontz, G.W. and Smith, L. S. (1968): Methods of using fish as biological research subjects. In *Methods of Animal Experimentation* (Ed. W. I. Gay), vol III, pp. 323-385. Academic Press, New York.

Konradt, A. G. (1968): Methods of breeding the grass carp, *Ctenophynogodon idella* and the silver carp, *Hypophythalmichthys molitrix*. FAO Fish Rep., 44(4): 195-204.

Kozlov, A.A. et al, (1977): Spravochnik po akklimatizatsii vodnykh organizmov (Bulletin on the acclimation of aquatic organisms). Moskva, Izd. Pishchevaya Promyshlennost, 174 p.

Krause, J. and Ruxton, G.D. (2002): *Living in groups*. Oxford, UK: Oxford University Press.

Kugino, K., Tamaru, S., Hisatomi, Y. and Sakaguchi, T. (2016): Long-Duration Carbon Dioxide Anesthesia of Fish Using Ultra Fine (Nano-Scale) Bubbles. Published online 11(4): doi: 10.1371/journal.pone.0153542

Kuhn, O. and Koecke, H.U. (1956): Histologische und cytologische Veranderungen der Fischkieme nach Einwirkung un Wasser enthaltener schadigender Substanzen. Zeitschrift fuer Zellforchung und Mikroskopische Anatomie., 43:611-643.

Kulkarni, C.V. and Ogale, S.N. (1979): The present status of mahseer (fish) and artificial propagation of *Tor khudree* (Sykes). J. Bomb. Nat. Hist. Soc., 75(3): 651-660.

Kumar, S. and Pant, S.C. (1981): Histopathologic effects of acutely toxic levels of Cu and Zn on gills, liver and kidney of *Puntius conchonius* (Ham.) Indian J. Exp. Biol. 19 : 191-194.

Kumari, V., Pandit, D.N., Shitanshu, S.K. and Singh, R.R. (2009): Anaesthetic effect of Benzocaine on oxygen consumption of a carp, *Labeo bata* (Ham). Trends Life Sci (India). 24(2): 25-29.

Kunwar, G.K., Pandey A. and Munshi, J.S.D. (1989): Oxygen uptake in relation to body weight of two freshwater major carps *Catla catla* (Ham.) and *Labeo rohita* (Ham.). J. Ind. I. Anim. Sci., 59(5): 621-624.

Laird, L.M. and Oswald, R.L. (1975): J. Inst. Fish Manage, 6(4): 92.

Lamholt, J.P. and Johansen, K. (1976): Gas exchange in the amphibious fish, *Ambhipnous cuchia*. J. Comp. Physiol., 107: 141-157.
Laming, P.R., Funston, C.W., Roberts, D. and Armstrong, M.J. (1982): Behavioural, physiological and morphological adaptations to the shanny (*Blennius pholis*) to the intertidal habitat. J. Mar. Biol. Assoc. 62:329-338.
Larmoyeux, J.D. and Piper, R.G. (1973): Effect of water reuse. Prog. Fish. Cult., 35: 2-8.
Lefevre, S., Bayley, M. and McKenzie, D.J. (2016): Measuring oxygen uptake in fishes with bimodal respiration. Journal of Fish Biology. 88, 206–231 doi:10.1111/jfb.12698,
Leis, O. (1978): Metod rascheta (Capacity calculation method) *Rybov. Rybolov,* (5):14–15.
Leitritz, E. and Lewis, R.C. (1976): Trout and salmon culture (Hatchery methods). Fish. Bull. Calif. Dep. Fish Game, (164):128–37
Lewis, S.V. (1979): A scanning electron microscope study of the gills of the air-breathing catfish, *Clarias batrachus*, L. J. Fish Biol., 15: 381–384.
Lighton, J.R.B. (2008). Measuring Metabolic Rates: A Manual for Scientists. New York, NY: Oxford University Press.
Lijuan, W., Shoukun, C, and Yinjiang, Z. (2015): The anesthetic effect of MS-222 in simulated transport of bass .24(2):235-241.
Lindstedt, Ph. (1914): Untresuch ungen uber Respiration Van Katt blutern. Zeitschrift fur Fisherei, 14:183.
Linton-Tyler, K., Reid, S.D. and Wood-Chris, M. (1998): The metabolic costs and physiological consequences to juvenile rainbow trout of a stimulated winter warming scenario in the presence or absence of sublethal ammonia. Trans. Am. Fish. Soc., 127(4): 611 – 619.
Lipskaya, N.Y. (1974): Metabolic rates in the young of some tropical fish species. J. Ichthyol., 14(6): 934-943.
Lizhen, L., Mingxi, D. and Lin, Y.L. (2019): Study on the Anesthesia Effect of MS-222 on *Pelteobagrus fulvidraco*. Aquatic Science and Technology Information. 46(1):62-66.
Lloyd, R. (1961): Effects of dissolved oxygen concentrations on the toxicity of several poisons to rainbow trout (*Salmo gairdneri*) Ann. Appl. Biol. 38: 447 – 455.
Lloyd, R. and Orr, L.D. (1969): The diuretic response by rainbow trout to sub-lethal concentrations of ammonia. Water Res. 3: 335–344.
Locke, D.O. (1969): Quinaldine as an anaesthetic for Brook trout, Lake trout and Atlantic salmon. Invest. Fish Control, No. 24, U. S. Sport Fish. Wildl., 5p.
Lusk, S. and Krcál, J. (1974): Preprava plôdika lososovitých rýb s kyslíkom (Transporting the fry of salmonids with oxygen.) Polov.Rybár. (2):32.
Maheshwari, U. K. (1990): Management of freshwater fish seed farms. Short-term training programme, CIFE 61–64.
Maki, A.W. (1979): Respiratory activity of fish as a predicator of chronic fish toxicity values for surfactants. In: Aquatic Toxicology. (Eds. L. L. Marking and R. A. Kimerle), Am. Soc. Test. Mat., pp. 77 – 95.
Malyukina, G.A., Aleksanchyuk, S.P. and Schtefanesku, M. (1962): On the role of the vision in "Schooling effect" in minnows *Phoxinus* L. and gold fish *Carassius carassius* L. Vopr. Ikhtiol, 2: 511 – 516.
Mammen, T.A. (1962): Live fish transport using modified splashless carriers with compressed aeration. In: Course manual of the training course on live fish transport. Hyderabad Fisheries Extension Unit, Govt. of India, 19-21 November, 1962.
Mammen, T.A. (1962b): Transport of freshwater fishes. In: Ibid. p4.
Mammen, T.A. (1967): Toxicity of MS-222 to selected fishes. U. S. Fish Wildl. Serv. Invest. Fish Control 18. 18p.
Mammen, T.A. (1962a): Live fish transport using modified splashless carrier with compressed in aeration. In: Training Course on Live Fish Transport, Nov. 19-21, Hyderabad. Fisheries Extension Unit, Government of India. P. 5.

Marking, L.L. (I967) Toxicity of MS-222 to selected fishes. US Fish Wildl. Serv. Invest. Fish Control 18. 18p.

Martyshev, F.G. (1983): *Pond Fisheries*. Amerind Publishing Co. Pvt. Ltd., New Delhi. pp 432-451.

Mashiko, K. and Jozuka, K. (1962): (In Japanese). Am. Rep. Notto. Mar. Lab., 2: 1 – 8.

Mathew, C.V. (1976): Oxygen consumption of *Mugil macrolepis* in relation to different salinities. Indian J. Fish., 23 II(1-2): 271-273.

Mc Donald, M. (1884): Report on the distribution of carp during the season of 1882. U. S. Comm Fish and Fisheries; Commissioners Report., 38: 915-942.

Mc Farland, W.N. (1959): A study of the effect of anaesthetics on the behaviour and physiology of fishes. Pub. Inst. Mar Sci., 6: 23-55.

Mc Farland, W.N. and Kolntz, G.W. (1969): Anaesthesia in fishes. Federation Proc. 28: 1535-1540.

McFarland, W.N. (1960): The use of anesthetics for the handling and transporting of fish. Calif. Dept. Fish Game., 66(4): 408 – 431.

Menon, A.G.K. (1974): A Checklist of the Fishes of the Himalayan and the Indogangetic Plains. F. S. I. Barrackpore.

Meyer, F.L. and Kramer, R.H. (1973): Effects of hatchery water reuse on rainbow trout metabolism. Prog. Fish. Cult., 35: 9 – 10.

Meyer, F.P., Schnick, R.A., Cumming, K.B. and Berger, B.L. (1976): Registration Status of fishery chemicals. Prog. Fish-Cult., 38(1): 3–7.

Miramontes, E., Mozdziak, P., Petitte, J.N., Kulus, M., Wieczorkiewicz, M. and Kempisty, B. (2020): Skeletal Muscle and the Effects of Ammonia Toxicity in Fish, Mammalian, and Avian Species: A Comparative Review Based on Molecular Research. Int. J. Mol. Sci. *21*(13): 4641. https://doi.org/10.3390/ijms21134641.

Mishra, B.K., Kumar, D. and Mishra, R. (1983): Observations on the use of carbonic acid anaesthesia in fish fry transport. Aquaculture, (3/4):405–8

Mitra, G.N. (1942): A new method of transport of fish fingerlings. Proc. Indian Sci. Cong., 3:159.

Mohapatra, B.C. and Saha, C. (2000): *Aquatic pollution and management.* Central Institute of Freshwater Aquaculture, Bhubaneswar, p. 1-363.

Moitra, A., Pandey, A.K. and Ghosh, T.K. (1979): Spawning behaviour of the climbing perch, *Anabas testudineus* (Bloch). Biol. Bull. India. 1(3): 31–34.

Morgan, M., Lumley, J., Whitwam, J.G. (1975): Etomidate, a new water soluble non-barbiturate intravenous induction agent – Lancet 7913: 955-956.

Morgulius, S. (1914): The body surface of flounders and its relation with gaseous metabolism. Am. J. Physiol., 36: 207 – 216.

Morris, R.W. (1967): High respiratory quotients of two species of bony fishes. Physiol. Zool. 40: 409-423.

Moss, D.D. and Scott, D.C. (1961): Dissolved oxygen requirements of three species of fish. Trans. Am. Fish. Soc., 90: 377-393.

Mosser, M.L. and Hettler, W.F. (1989): Routine metabolism of juvenile sport, *Leiostormus xanthuerus* (Laccpedle), as a function of temperature, salinity and weight. J. Fish Biol., 35: 703-707.

Mottias, A.T., Rantin, F.I. and Fernandes, M.N. (1998): Gill respiratory parameters during progressive hypoxia in the facultative air–breathing fish, *Hypostomus rgani* (Loricariidae). Comp. Biochem. Physiol., 120(2): 311– 315.

Motwani, M.P. and Bose, B.B. (1957): Oxygen requirements of fry of the Indian major carp, *Labeo rohita* (Hamilton) Proc. Nat. Inst. Sci. Indian. 23(B): 8– 16.

Moyle, P.B. and Cech, J.J. (2003): *Fishes, An Introduction to Ichthyology.* 5th Ed, Benjamin Cummings.

Mulligan, H.W. and Majid, A. (1936): REC. Malaria Surv. India, 6(4): 537–547.

Munshi, J.S.D., Ojha, J., Ghosh, T.K., Roy, P.K. and Mishra, A.K. (1984): Scanning electron microscopic observations on the structure of gill rakers of some freshwater teleostean fishes. Proc. Indian Natn. Sci. Acad., B50: 549-554.

Munshi, J.S.D. and Singh, A. (1992): Scanning electron microscopic evaluation of effects of low pH on gills of *Channa punctatus* (Bloch). J. Freshwater Biol., 4(1): 83–89.

Munshi, J.S.D. and Ghosh, T.K. (1993): Metabolic wheel hypothesis as applied air –breathing fishes of India. In "*Advantage in fish Biology*". (Ed. H. R. Singh). Pp. 70 – 78. Hindustan Publishing Corporation (India), Delhi.

Munshi, J.S.D., Choubey, B.J., Pandey, P.K. and Pandey, B.N. (1977): Effect of hydrocortisone and L-thyroxine on the blood volume and haematocrit values of an air – breathing fish *Heteropneustes fossils* (Bloch). Zoologische Beitrage, 23: 273-280.

Munshi, J.S.D., Patra, A.K. and Hughes, G.M. (1982): Oxygen consumption from air and water in *Heteropneustes (=Saccobranchus) fossilis* (Bloch) in relation to body weight at three different seasons. Proc. Indian natl. Sci. Acad., B48(6): 715 – 729.

Munshi, J.S.D., Patra, A.K., Biswas, N. and Ojha, J. (1979): Interspecific variations in the circadian rhythm of bimodal oxygen uptake in four species of Murrels. Jap. J. Ichthyol., 26 (1): 69-74.

Munshi, J.S. D., Sinha, A. L. and Ojha, J. (1976): Oxygen uptake capacity of gills and skin in relation to body weight of the air-breathing fish, *Clarias batrachus*. Acta. Physiol. Acad. Sci. Hungaricae., 48: 28-33.

Munshi, J.S.D. and Dube, S.C. (1973): Oxygen uptake capacity of gills in relation to body size of the air-breathing fish, *Anabas testudineus* (Bloch). Acta. Physiol. Acad. Sci. Hungaricae., 44: 113 – 123.

Munshi, J.S.D. and Hughes, G.M. (1992): *Air-breathing Fish: Their structure, function and life history*. Oxford and IBH Pub. Co., New Delhi p. 255.

Munshi, J.S.D., Pandey, B. N., Pandey, P.K. and Ojha J. (1978): Oxygen uptake through gills and skin in relation to body weight of an Air-breathing siluroid fish, *Saccobranchus (=Heteropneustes) fossilis* (Bloch). J. Zool. Lond. 184: 171-180.

Mylonas, C.C., Cardinaletti, G., Sigelaki, I. and Polzonetti Magni, A. (2005): Comparative efficacy of clove oil and 2-phenoxyethanol as anesthetics in the aquaculture of European sea bass (*Dicentrarchus labrax*) and gilthead sea bream (*Sparus aurata*) at different temperatures. Aquaculture. 246(1): 467-481.

Nadler, L.E., Killen, S.S., McClure, E.C., Munday, P.L. and McCormick, M.I. (2016): Shoaling reduces metabolic rate in a gregarious coral reef fish species. J. Exp. Biol. 219: 2802–2805.

Narsimham, C. and Parvatheswararao, V. (1974): Adaptation to osmotic stress in a freshwater euryhaline teleost, *Tilapia mossambica*. X. Role of mucopolysaccharides. Acta histochem., 51: 37-49.

Natarajan, M.V. and Ranganathan, V. (1960): A note on the possibilities of utilizing quinaldine in transporting live fish. Curr. Sci., 29: 393.

Nath, B. (1966): Prehistoric animals of India and their bearing on early Indian culture, Proc. All India Cong. Zool., 2(2): 6-13.

Neelima, P, Gopala Rao, N., Srinivasa Rao, G. and Chandra Sekhara Rao, J. (2016): A Study on Oxygen Consumption in a Freshwater Fish *Cyprinus carpio* Exposed to Lethal and Sublethal Concentrations of Cypermethrin (25%Ec). Int. J. Curr. Microbiol. App. Sci. 5(4): 338-348.

Nekvasil, N.P. and Olson, K.R. (1986): Extraction and metabolism of circulating catecholamines by the trout gill. Amer. J. Physiol., 250 (Regul. Integr. Comp. Physiol. 19), R526-531.

Nieminen, M., Laitinen, M. and Pasanen, P. (1982): Effects of anaesthesia with Tricaine (MS_{222}) on the blood composition of the splake (*Salvelinus nails x Salvelinus mamaycush*). Comp Biochem. Physiol., 73C: 271-276.

Norman, J.R. (1975): Chapter 5: Respiration. In *A History of fishes*. (Ed. P.H. Greenwood) Pub. Ernest Benn Limited London. Pp. 76-96.

Ojha, J and Haque, Z. (1994): Changes of scaling pattern in oxygen uptake in relation to body weight of an estuarine goby, *Pseudapocryptes lanceolatus* (Gobiidae, Perciformes). *Proc. Indian ntn. Sci. Aca.* B60 (6): 505–512.

Ojha, J. and Singh, R. (1981): Oxygen uptake in relation to body weight of the freshwater catfish, *Mystus cavasius* (Ham.). Indian. J. Exp. Biol., 19(2): 126-129.

Ojha, J. and Munshi J.S.D. (1975): Oxygen consumption in relation to body size and respiratory surface area of a freshwater mud eel, *Macrognathus aculeatum* (Bloch). Indian J. Exp. Biol., 13: 353-357.

Ojha, J., Dandotia, O.P. and Munshi J.S.D. (1977): Oxygen consumption of an amphibious fish, *Colisa fasciatus* in relation to body weight. Pol. Arch. Hydrobiol., 24(4): 547-553.

Ojha, J., Dandotia, O.P., Patra, A.K., and Munshi, J.S.D. (1978): Bimodal oxygen uptake in relation to body weight in a freshwater murrel, *Channa (=Ophiocephalus) gachua*. Z. Tierphysiol., 40: 57-66.

Ojha, J., Mishra, N., Saha, M.P. and Munshi, J.S.D. (1979): Bimodal oxygen uptake in juveniles and adults amphibious fish *Channa (=ophiceplalus) marulius*. Hydrobiol., 63 (2): 153 – 159.

Ojha, J., Patra, A.K., Biswas, N. and Munshi, J.S.D. (1981): Interspecific variations in the bimodal oxygen uptake of the genus *Channa*. Z. Tierphysiol., Tierernahrg. u. Futtermittelkde., 46: 165-170.

Ojha, J., Rooj, N.C., Mittal, A.K. and Munshi, J.S.D. (1989): Light and scanning electron microscopic studies on the effect of biocidal plant sap on the gills of a hill-stream fish, *Gara lamta* (Ham.). J. Fish. Biol., 34: 165-170

Ojha, J., Singh, N.K. and Ghosh, T.K. (1988): Bimodal oxygen uptake in relation to body weight in a freshwater air-breathing fish, *Colisa (=Trichogaster) fasiatus* (Bloch and Schneider). Arch. Hydrobiol., 111(3): 475-480.

Olson, K.R. and Fromm, P.O. (1973): A scanning electron microscopic study of secondary lamellae and chloride cells of rainbow trout (*Salmo gairdneri*). Z. Z. Zellforsch., 143: 439-449.

Olson, K.R., Kullman, D., Narkates, A.J. and Oparil, S. (1986): Angiotensin extraction by trout tissues in vivo and metabolism by the perfused gill. Amer. I. Physiol. 250 (Regul. Intefr. Comp. Physiol. 19), R532–R538.

Olson, K.R. and Fromm, P.O. (1971): Excretion of urea by two teleosts exposed to different concentrations of ambient ammonia. Comp. Biochem. Physiol., 40A: 999–1007.

Orlov, Yu.I. et al., (1974): Transportirovka zhivoi ryby v germeticheskikh emkostyakh. *Spravochnoe posobie* (Live fish transport in hermetically sealed containers. Information manual) Moskva, Izd. Pishchevaya Promyshlennost', 97p.

Osborn, P.E. (1951): Some experiments on the use of thiouracil as an aid in holding and transporting fish. Prog. Fish. Cult., 13:75-78.

Ostrensky, A. and Brugger, A. (1992): Studies on the viability of silverside *Odontesthes argentinensis* cultivations: acute toxicity of ammonia. Ciências e Cultura., 44: 413–414.

Ou-youjum and Li-Jia'-Er (1996): Test on the transportation of live parent mullet. J. Mar. Fish. – Haiyang- Yaye, 18(1): 8-12.

Packard, M.J. and Stevensen, R.K. (1974): The influence of thyroxin on oxygen consumption of tissues from the frog, *Rana pipiens*. Gen Comp. Endocrinol., 22(2): 195 – 198.

Packer, R. and Dunson, W. (1970): Effect of low environmental pH on blood pH and sodium balance of brook trout. J. Exp. Zool., 174: 65 – 72.

Paley, R.K. Twitchen, I.D. and Eddy, F.B. (1993): Ammonia, Na^+, K^+, and Cl^- levels in rainbow trout yolk sac fry in response to external ammonia, J. Exp. Biol., 180: 273 – 284.

Paloheimo, J.E. and Dickie, L.M. (1965): Food and growth of fishes. II. Effects of food and temperature on the relation between metabolism and body weight. J. Fish. Res. Bd. Can., 23: 869-908.

Pandey, A. (1988): *The structure and function of the respiratory organs of major carp, Labeo rohita (Ham.)*. Ph. D. thesis of Bhagalpur University, Bhagalpur, India.

Pandey, B.N. (1975): Effect of salinity and low pH on the blood and oxygen consumption in an air breathing fish, *Heteropneustes fossilis* (Bloch). J. Indian Biosci. Assoc., 1: 137 – 140.

Pandey, B.N., Chanchal, A.K. and Singh, M.P. (1976): Effect of malathion on oxygen consumption and blood of *Channa punctatus*, Indian J. Zoot., 16: 95-100.

Pandey, B.N., and Munshi, J.S.D. (1976): Role of thyroid gland in regulation of metabolic rate in air breathing siluroid fish, *Heteropneustes fossilis* (Bloch). J. Endocrinol., 69: 421-425.

Pandit, D.N. and Ghosh, T.K. (2005): Effect of Benzocaine on opercular movement and oxygen consumption in *Channa punctatus* (Bloch). J Indian Fish Assoc. 32:149-155.

Pandit, D.N. and Ghosh, T.K. (2006): A comparative study on oxygen uptake during still water aggregation of *Anabas testudineus* and *Channa punctatus*. Modern J Life Sci. 5(1-2): 7-12.

Pandit, D.N. and Ghosh, T.K. (2007): Oxygen uptake in relation to group size in the juveniles of a climbing perch, *Anabas testudineus*. J Environ Biol. 28(1): 141-143.

Pandit, D.N. and Ghosh, T.K. (2012): Optimization of packing conditions for live transportation of *Anabas testudineus* and *Channa punctatus*. Fish Technol. 49: 59-63.

Pandit, D.N., Singh, K.N., Kumari, D. and Sharma, S.K. (2008): Oxygen uptake in *Channa punctatus* (Bloch) from pond water versus tap water: A requirement of packing of live fish. J Curr Sci. 12(2): 619-623.

Pandit, D.N., Singh, K.N., Kumari, D. and Sharma, S.K. (2009): Comparison of oxygen uptake of *Channa punctatus* (Bloch) from tube-well water and tap water: Bull Environ Sci. 28(2): 143-148.

Pandit, D.N. and Ghosh, T.K. (1999): Effect of an anaesthetic, Benzocaine on the aquatic oxygen uptake in the juveniles of facultative air-breathing fish, *Heteropneustes (=Saccobranchus) fossilis* (Bloch). J. Freshwater Biol. 12(3-4):45-48.

Parker, F. R. (1973): Reduced Metabolic Rate in fishes as a Result of induced schooling. Trans. Am. Fish. Soc., 102:125–131.

Parvatheswararao, V. (1959): Studies on the oxygen consumption in tropical poikilotherms. III. Oxygen consumption in the freshwater fish *Etroplus maculates* (Bloch) in relation to size and temperature. J. Anim. Morph. Physiol., 4:34-37.

Parvathy, U., Kumar, K.S., Binsi, P.K., Nambiar, L., Ninan, G. and Zynudheen, A.A. (2019): Effect of Anaesthetics, Temperature and Aeration in Live Transportation of Tilapia (*Oreochromis mossambicus*) (Peters, 1852). Fish Tech. 56: 38-43.

Pathak, P., Pandey, P.K., Thakur, P.K. and Choudhary, B.P. (1992): Effect of salinity on endocrine organs, corpuscular haematology and rate of oxygen uptake in *Channa punctatus* (Bloch). J. Curr. Biosci., 9(2): 49 – 52.

Patra, A.K., Biswas, N., Ojha, J. and Munshi. J.S.D. (1978): Circadian rhythm in bimodal oxygen uptake in an obligatory air-breathing swamp eel *Amphipnous (=Monopterus) cuchia* (Ham). Ind. J. Exp.Biol., 16 (7): 808-809.

Patra, A.K., Munshi, J.S.D. and Hughes, G.M. (1983): Oxygen consumption of the freshwater air-breathing Indian siluroid fish, *Clarias batrachus* (Linn.) in relation to body size and seasons. Proc. Ind. Natn. Sci. Acad., 49 (6): 566-574.

Patro, J. C., (1968): A note on a new fish-carrier, double barrel type (mimeo): p6.

Paul, V.I. and Banerjee, T.K. (1995): Acute toxicity of ammonium sulphate to the breathing organ of the live fish *Heteropneustes (Saccobranchus) fossilis* (Bloch). Curr. Sci., 68: 845-849.

Paul, V.I. and Banerjee, T.K. (1996): Ammonium sulphate induced stress related alteration in the respiratory epithelium of the air-breathing organ of the catfish *Heteropneustes fossilis* (Bloch). J. Biosci., 21: 519 – 526.

Pecha, O.R., Berka R. and Kouril, J. (1983): Preprava pludku v polyetylénovych vacích (Fry transport in polyethylene bags) Ser.Metod.VURH Vodnany, (10):16 p.

Pelster, B., Wood, C.M., Campos, D.F. and Adalberto, L.V. (2020): Cellular oxygen consumption, ROS production and ROS defence in two different size-classes of an Amazonian obligate air-breathing fish (*Arapaima gigas*). Plos One. https://doi.org/10.1371/journal.pone.0236507

Pelster, B., Bridges, C.R. and Grieshaber, M.K. (1988): Physiological adaptations of the intertidal rockpool teleost, *Blennius pholis* L., to aerial exposure. Respir. Physiol. 71: 355-374.

Pelster, B., Wood, C.M., Campos, D.F. and Adalberto, L.V. (2020): Cellular oxygen consumption, ROS production and ROS defence in two different size-classes of an Amazonian obligate air-breathing fish (*Arapaima gigas*). Plos One. https://doi.org/10.1371/journal.pone.0236507

Perry, S.F. and Laurent, P. (1989): Adaptational response of rainbow trout to lowered external NaCl concentration: contribution of the branchial chloride cell. J. Exp. Biol., 147: 147–168.

Peters, G., Hoffmann, R. and Klinger, H. (1984): Environment-induced gill disease of cultured rainbow trout (*Salmo gairdneri*). Aquacult. 38: 105 – 126.

Peterson-Curtis, T.L. (1997): Effect of salinity on survival, growth, metabolism and behaviour in juvenile hogchokers, *Trinectes maculates fasciatus* (Archiridae). Environ. Biol. Fishes., 49: 323 – 331.

Piper, R.G. et al., (1982): Fish hatchery management. Washington, D.C., U.S. Department of the Interior, Fish and Wildl. Service, pp. 348–71

Pisam, M., Boefu, B., Prunet, P. and Ramburg. A. (1990): Ultrastructural features of mitochondria rich cells in stenohaline freshwater and seawater fishes. Am. J. Anat., 187: 24 – 31.

Pitcher, T.J. (1975): A Periscopic Method for Determining the Three – Dimensional Positions of Fish in School. J. Fish. Res. Bd. Can., 32: 1533 – 1538.

Prama de Croux, M.J. (1997): Relationship between metabolic rate and body weight in *Salminus maxillosus* (Valenciennes, 1840). J. Aqua. Trop., 12(1): 17 – 22.

Prigogine, I. and Wiame, J.M. (1946): Biologic el thermadynamique des phenomenas irreversible. Experimentia 2: 451-453.

Privol'nev, T.I. (1964): Vyzhivanie i perevozka zhivoi ryby bez vody vo vlazhnoi atmosphere (Survival and waterless transportation of live fish in humid weather). Izv. Gos N10RKh. Leningrad. vol. 58.

Proske, Ch. (1982): Uberlandfahrt im kühlen Wasser. Der Transport von Fishen bringt im Sommer eine Menge Probleme. Landwirtsch.Wochenbl., 172(15):11–12.

Prosser, C.L. (1961): Respiratory function of the body fluids. In *Comparative Animal Physiology*. W.B. Saunders and Co., Philadelphia and London.

Rahn, H. (1966): Evolution of gas exchange in vertebrates. In "Studies in pulmonary Physiology, Mechanics, Chemistry and Circulation of the lungs". Aerospace Med. Res. Lab. 2: 42 – 54.

Raiesi, S.N., Mazandarani, M., Shahidi, S-A., Azade Saraei, A.G-H., Khani, F. and Fatahi, S. (2019): Acute and chronic toxicity of ammonia in Persian sturgeon, *Acipenser persicus,* fingerlings. International Journal of Aquatic Biology

Raj, B.S. and Cornelius, R.P. (1949): Transport of mirror carp from Ootacamund to Kumaon Hills. Indian Farm., 10: 56 – 58.

Rajbanshi, V. (1977): The architecture of the gill surface of the catfish, *Heteropneustes fossilis* (Bloch), SEM study. J. Fish. Biol., 10: 325 – 329.

Ram, R.N. and Sathyanesan, A.G. (1987a): Effect of chronic exposure of commercial nitrogenous fertilizer, ammonium sulphate on testicular development of a teleost, *Channa punctatus* (Bloch). Ind. J. Exp. Biol., 25(10): 667-670.

Ram, R.N., and Sathyanesan, A.G. (1986): Ammonium Sulphate induced nuclear changes in the oocyte of the fish, *Channa punctatus* (Bloch). Bull. Environ. Contam. Toxicol., 36: 871 – 875.

Ram, R.N., and Sathyanesan, A.G. (1987b): Histopathological changes in liver and thyroid of the teleost fish *Channa punctatus* (Bloch) in response to ammonium sulphate fertilizer treatment. Ecotoxicol. Environ. Safety., 13:185-190.

Ramachandran, V. (1972): Short and long distance transport of live fish for stocking. Souvenir, Silver Jubilee CIFRI, Barrackpore, p. 45-48

Ramaswamy, L.S. and Sundararaj, B.I. (1957): Induced spawning in the catfish, *Clarias*. Naturwissenchaften 44: 384.

Ramchandran, V. (1969): Transport of spawn, fry, fingerlings and breeders. FAO/ UNDP Regional Seminar on Induced Breeding of Cultivated Fishes. Calcutta-Cuttak-Bombay. FRI/IBCF 16: 28.

Ranade, M.R. and Kewalramani, H.G. (1956): Experimental air-shipment of carp fry in plastic bags. Ind. J. Fish., 3(2):290-293.

Randall, D.J., Perry, S.F. and Heming, T.A. (1982): Gas transfer and acid/ base regulation in salmonids. Comp. Biochem. Physiol., 73B: 93 – 103.

Ranjeet, K., Divya, D.K., Fathimath, S.M.P. and Reshmi, K. (2013): Effect of Ekalux Toxicity on Selected Physiological Parameters In *Anabas testudineus*. Applied Ecology and Environmental Research 11(4): 569-580.

Ranke-Rybicka, B. (1975): Effect of organophosphate pesticides (Malathion, Foschlor, Dichlorophos) on aquatic organism. ROCZN. PZH., 26: 393 – 399.

Reddy, K.P.O., Latha, E.S., Muthuraj, D. and Neeraja, P. (1996): Certain metabolic modulations in carbohydrate metabolism of fry, *Cyprinus carpio* on ammonia stress. J. Inland Fish. Soc. India., 28(2): 42 – 45.

Reddy, T.G., Palaniappan, S. And Pillay, K.B.P. (1977): Respiratory and histopathological effects of the pesticide Disyston on the fish, *Colisa latia*. Environ. Bio., p. 57.

Reichenbach-Klinke, H.H. (1967): Untersuchungen uber die Einwirkung des Ammoniakgehaltes auf den Fischor ganismus. Arch. Fischerei. Wiss. 17: 122.

Remen, M., Imsland, A.K., Stefansson, S.O., Jonassen, T.M. and Foss, A. (2008): Interactive effects of ammonia and oxygen on growth and physiological status of juvenile Atlantic cod (*Gadus morhua*). Aquaculture, 274(2-4): 292-299.

Rhyzhkov, L.P. (1970): The role of certain receptors in the formation of the group effect in the bream, *Abramis brama* (L.). J. Ichthyol., 10: 358 – 363.

Richardson, J. (1997): Acute ammonia toxicity for eight New Zealand indigenous freshwater species. NZ. J. Mar. Freshwater Res. 31: 185–190.

Robinette, H.R. (1976): Effects of sublethal level of ammonia on the growth of channel catfish. Arch. Fisherei Wiss., 38: 26 – 29.

Rodriguez-Gutierrez, M. and Esquirvel-Herrera, A. (1995): Aquaculture. 129 – 431.

Rombough–Peter–J. (1988): Partitioning of oxygen uptake between gills and skin of fish larvae. A novel method for estimating cutaneous oxygen uptake. J. Exp. Biol, 201(11): 1763-1769.

Ross, L.G. and Ross, B. (1984): Anaesthetic and Sedative Technique for Fish, Institute of Aquaculture, University of Stirling, Scotland, p.35.

Roy, P.K. and Munshi, J.S.D. (1988): Oxygen consumption and ventilation rate of a freshwater major carp, *Cirrhinus mrigala* (Ham.) in fresh and malathion treated waters. J. Environ. Biol., 9: 5 – 13.

Roy, P.K. and Munshi, J.S.D. (1991): Malathion induced structural and morphometric changes of gills of a freshwater major carp, *Cirrhinus mrigala* (Ham.) J. Environ. Biol., 12(1): 79 – 87.

Roy, P.K., Munshi, J.S.D. and Munshi, J.D. (1986): Scanning electron microscopic evaluation of effect of saponin on gills of the climbing perch, *Anabas testudineus* (Bloch). (Anabantidae; Pisces). Indian J. Expt. Biol. 24: 511–516.

Roy, P.K., Munshi, J.S.D. and Dutta, H.M. (1990): Effect of Saponin extracts on morpho histology and respiratory physiology of an air-breathing fish, *Heteropnuestes fossilis* (Bloch). J. Freshwater Biol., 2(2): 135 - 145

Roy, P.K. and Munshi, J.S.D. (1984): Oxygen uptake in relation to body weight and respiretory surface area in *Cirrhinus mrigala* (Ham.) at two different seasonal temperatures. Proc. Indian. Nat. Sci. Acad. B 50(4): 387-394.

Rubner, M. (1883): Uber den einfluss der korpergrosse auf. Stoff-und kraftwechsel. Z. Biol., 19: 535 – 562.

Russo, R.C. and Thurston, R.V. (1991): Toxicity of ammonia, nitrite and nitrate to fishes. In: *Aquaculture and Water Quality* (Brune E., Tomasso J.R.), World Aquacultural Society Publication, pp. 58-89.

Ruttner, F. (1926): Bermerkungen uber Sanerstoffgehalt der Gewasser und desson respiratorischen Wert. Naturevissenschaften Bd. 14(5): 1237– 1239.

Saha, N. and Ratha, B.K. (1994): Induction of ornithineurea cycle in a freshwater teleost, *Heteropneustes fossilis*, exposed to high concentrations of ammonium chloride. Comp. Biochem. Physiol. 108: 315-325

Saha, G.N. (1966): A note on the oxygen requirement of spawn of Indian major carps. Sci. Cult., 32(6): 327–328.

Saha, K.C. and Sen, D.P. (1958): Studies on the mortality in spawn and fry of Indian major carps during transport. Part–IV. Types of containers used for transporting spawn and fry under oxygen pressure. Ind. J. Fish. 5(2): 362–374.

Saha, K.C., Sen D.P. and Majumdar P. (1956b): Studies on the mortality in spawn and fry of Indian major carps during transport. Part III. Effect of inimical substances from decomposition of metabolic products in the medium on spawn life and their control. Ibid., 3(1):135–40.

Saha, K.C., Sen, D.P. and Banergea, S.M. (1955): Use of sodium amytal in the transport of fry of Indian major carps. Ann. Biochem., 15(2): 133.

Saha, K.C., Sen, D.P. and Majumdar P. (1956a): Studies on the mortality in spawn and fry of Indian major carps during transport. Part II. Effect of oxygen pressure, free surface area, water volume and number of fry in the medium of transport. Indian J. Fish., 3(1): 127 – 134.

Saha, K.C. and Chowdhury N.K. (1956): Studies on the mortality in spawn and fry of Indian major carps during transport. Part – I. Investigations relating to the use of specific soil (red soil) - water medium in transporting carp spawn ('Dimpona') in the existing trade to check mortality during transit. Ind. J.Fish., 3(1):119–26.

Santhi, K., Gunasekhar, K., Babur, C.S., Reddy, K.P.O. and Neeraja, P. (1992): Changes in the dehydrogenase levels of fish *Orechromis mossambicus* under chronic ammonia stress. J. Inland Fish. Soc. India, 24(2): 71 – 73.

Sarkar, S.K. and Pramanik, A. (1986): Evaluation of the toxicity of two nitrogenous compounds on freshwater invertebrates exposed to temperature. J. Environ. Biol. 7: 183 – 188.

Satyanarayana, V. (1996): *A Symposium on Fish Culture*. Narendra Publishing House, New Delhi. 234p.

Saunders, R.L. (1963): Respiration in the Atlantic Cod. J. Fish. Res. Bd. Can., 20:373 – 386.

Scgoettger, R.A. and Steucke, E.W. (1970): Synergic mixtures of MS-222 and quinaldine as anaesthetics for channel catfish: Its toxicity, efficacy and muscle residues. Fish Control, No.17, I. B. Sport Fish. Wildl., pp. 1-14.

Schaeperclaus, W. (1933): *Text- book of pond – culture*. Translated by E. Huud as *U. S. Fish and Wild Life Service*. pp. 1–113.

Schimidt-Neilson, K. (1984): *Scaling: why is animal size so important?* Cambridge University Press, Cambridge.

Schoettger, R.A. and Julin, A.M. (1967): Effects of MS_{222} as an anesthetic on four salmonids. U.S. Fish Will. Serv. Invest Fish Control., p. 15.

Scholander, P.F., Flagg, W., Walters, V. and Irving, L. (1953): Climatic adaptation in arctic and tropical poikilotherms. Physiol. Zool., 26: 67 – 82.

Schtefanesku, M. (1958): Physiological analysis of the correlation between gas exchange and schooling of some marine and freshwater fish. Zool. Zh., 34(2): 222 – 228.

Schuett, F. (1934): Studies in mass physiology: the activity of goldfish under different conditions of aggression. Ecology, 15: 158 – 262.

Schulze-Wiehenbrauk, H. (1976): Effects of sulethal ammonia concentrations on metabolism in juvenile trout (*salmo gairdeneri* Rich). Ber. Ostch. Wiss. Komm. Meeresforsch., 24: 234 – 256.

Schurmann, H. and Steffensen, J.F. (1997): Effects of temperature, hypoxia and activity on the metabolism of juvenile Atlantic cod. J. Fish. Biol., 50: 1166 – 1180.

Scott, G.R., Matey, V., Mendoza, J.A., Gilmour, K.M., Perry, S.F. and Almeida-Val, V.M.F. (2017): Air breathing and aquatic gas exchange during hypoxia in armoured catfish. J Comp Physiol B. 187(1): 117–133. Available from: pmid: 27461227.

Selvaraj, C., Mohanti, A.N. and Ghosh, S.R. (1981): Preliminary observations on the transport of fish fry under air pressure. J. Inland Fish. Soc. India., 13(2): 75 – 79.

Sharma, S.K., Durve V.S. and Sharma, L.L. (1991): Fishing Chimes, 39: 38.

Shaw, E. (1970): Schooling in fishes: Crtique and review. In: *Development and evolution of behaviour* (Eds. Aronson, L. R., Tobach, E., Lehrma, D. S., Resenblatt, J. S. and Resenblatt, J. S.). pp. 452 – 480. San Francisco,

Shivraj, K.M. and Patil, H.S. (1985): Acute toxicity of Cd and Co to a freshwater fish *Lepidocephalichthys guentea*. Indian J. Comp. Anim. Physiol., 3: 24 – 28.

Shlaifer, A. (1938): Studies in mass physiology: effects of numbers upon the oxygen consumption and locomotor activity of *Carrassius auratus*, Physiol. Zool., 11(4): 408 – 424.

Shlaifer, A. (1939): An analysis of the effect of numbers upon the oxygen consumption of *Carassius auratus*. Physiol. Zool., 12(4): 381 – 392.

Shtefan, M. (1958): Physiological analysis ofthe interrelationships between the gas exchange and shoal behaviour of certain marine and freshwater fishes. Zoollogicheskii Zhurnal, 37:222–228.

Singh, A. (1990): *Histophysiology of respiratory organs and skin of normal and heavy metal and acid exposed fishes*. Ph. D. Thesis, Bhagalpur University, Bhagalpur. (India). 185p.

Singh, A. and Munshi, J.S.D. (1996): Scanning electron microscopic evaluation of the effects of short and long term exposure of copper and mercury on the gills of *Channa punctatus* (Bloch.). J. Freshwater Biol. 8(1): 51–55.

Singh, B. N. and Hughes, G.M. (1973): Cardiac and respiratory responses in the climbing perch *Anabas testudineus.* J. Comp. Physiol. 84:205-226.

Singh, B.N. (1976): Balance between aquatic and aerial respiration. In *Respiration of Amphibious Vertebrates* (Ed. G. M. Hughes), pp.125–164. Academic Press, London.

Singh, B.N. (1977): Oxygen consumption and the amount of oxygen required for transport of rohu and mrigala fingerlings. J. Inland Fish. Soc. India., 9: 98-104.

Singh, B.N. and Hughes, G.M. (1971): Respiration in an air-breathing catfish, *Clarias batrachus* (Linn.). J. Exp. Biol., 55: 421–434.

Singh, B.N. and Munshi, J.S.D. (1985): Oxygen uptake in relation to body weight, respiratory surface area and group size in a freshwater goby, *Glossogobius giuris* (Ham.). Proc. Indian natn. Sci. Acad., B 51(1): 33-40.

Singh, S.B., Ghosh, S.R., Dey, R.K., Reddy, P.V.G.K. and Mishra, B.K. (1979): Oxygen packing of silver carp and grass carp fingerlings at different densities and temperature ranges. In Symposium on Inland Aquaculture (Abstract), CIFRI, Barrackpore. 9: 12 – 14.

Singh, S.R. and Singh, B.R. (1979): Changes in oxygen consumption of a siluroid fish (*Mystus vittatus*) put to different concentrations of some heavy metal salts. Indian J. Exp. Biol. 17(3): 274 – 276.

Sinha, M.K., Adhikari, S., Roy, P.K., Munshi, J.D. and Munshi, J.S.D. (1997): Scanning Electron Microscopic (SEM) Evaluation of the Effects of *Strychnos nux-vomica* on Gills of Indian minor carp Labeo bata (Ham.). J. Freshwater Biol., 9(1): 37 – 46.

Siwicki, A. (1984): New anaesthetic for fish – Aquaculture 38: 171-176.

Smart, G.R. (1975): The acute toxic mechanisms of ammonia to rainbow trout (*Salmo galrdneri*). Ph D. thesis, University of Bristol, England.

Smart, G.R. (1976): The effect of ammonia exposure on gill structure of the rainbow trout (*Salmo gairdneri*). J. Fish Biol. 8: 471–475.

Smart, G.R. (1978): Investigations of the toxic mechanisms of ammonia to fish-gas exchange in rainbow trout (*Salmo gairdneri*) exposed to acutely lethal concentrations. J. Fish. Biol. 12: 93–104.

Smatresk, N.J., and Herreid, C.F. (1980): Group metabolism in swordtails, *Xiphophorus helleri*, under controlled oxygen conditions. Copeia, 562–564.

Smit, G.L., Barham., W.T. and Schoonbee, H.J. (1977): Some effects of the anaesthetic MS–222 on freshwater. S. Af. J. Sci., 73:351 – 352.

Smit, G.L., Hattingh, J. and Burger, A. P. (1970): Haematological assessment of the effects of the anaesthetics MS_{222} in natural and neutralized form in three freshwater fish species: intra-specific differences. J. Fish. Biol., 15: 645 – 653.

Smith, H.W. (1935a): The metabolism of lungfish I. General consideration of the fasting metabolism on active fish. J. Cell. Comp. Physiol. 6: 43-67.

Smith, H.W. (1935b): The metabolism of lung fish II. Effect of feeding meat on the metabolism rate. J. Cell. Comp. Physiol. 6: 335-349.

Smith, L.S. and Bell, G.R. (1967): Anesthetic and surgical techniques for pacific salmon. J. Fish. Res. Bd. Can., 24: 1579 – 1588.

Smits, J. and Permanyer, I. (2019): The sub-national human development database. Sci. Data 6, 190038.

Snow, J.R., Brewer, D. and Wright, C.F. (1978): Plastic bags for shipping sac fry of largemouth bass. Progr. Fish-Cult. 40(1): 13–4

Soderberg, R.W., Flynn, J.B. and Schmittou, H.R. (1983): Effects of ammonia on growth and survival of rainbow trout in intensive static-water culture. Trans. Am. Fish. Soc. 112: 448 – 451.

Soderberg, R.W., McGee, M.V., Grizzle, J.M. and Boyd, C.E. (1984): Comparative histology of rainbow trout and channel catfish grown in intensive static water aquaculture. The Prog. Fish-Cult., 46: 195–199.

Sousa, R.J , Meade, T.L. (1977): The influence of ammonia on the oxygen delivery system of coho salmon hemoglobin. Comp. Biochem. Physiol. 58 A: 23 – 58.

Sperry, D.G. and Wassersug, R.J. (1976): A proposed function for micro ridges on epithelial cells. Anat. Rec., 185: 235 – 258.

Sreenivasan, A. (1962): Use of tranquilizing drug and narcotics in reducing oxygen consumption by fish. Indian J. Fish., 9A(2): 738 – 745.

Srivastava, C.B. and Karamchandani, S.J. (1964): Conditioning of carp fry for transportation. Indian J. Fish., 11(2): 503 – 520.

Stokesa, G.L., Abigail, J. Lynchb, Benjamin, S. Lowea, Simon Funge-Smithc, John Valbo Jørgensenc, and Samuel, J.S. (2020): COVID-19 pandemic impacts on global inland fisheries. PNAS. 117(47): 29419-29421. http://www.pnas.org/cgi/doi/10.1073/pnas.2014016117

Strange, R.J. and Schreck, C.B. (1978): Anesthetic and handling stress on survival and cortisol concentration in yearling chinook salmon (*Oncorhynchus tshawytscha*). J. Fish. Res. Bd. Can. 35: 345–349.

Strebkova, T.P. (1971): Ispol'zovanie narkoticheskoi smesi pri transportirovke zhivoi ryby (The use of a narcotic mixture in the transport of live fish). Rybn. Khoz. (12):19–20.

Strzyzewska, E., Szarek, J. and Babinska, I. (2016): Morphologic evaluation of the gills as a tool in the diagnostics of pathological conditions in fish and pollution in the aquatic environment: a review. *Veterinarni Medicina,* 61: (3): 123–132

Studnicka, M. et al, (1982): Zastosowanie środka antystresowego w transporcie pstraga (The use of an anti-stress preparation in the transport of trout). Gospod. Rybn. 34(3):10–11.

Subba, B.R. (1999): *Structure and function of the respiratory organs of a hill stream fish, Glyptothorax telchitta* (Ham.) from Nepal. Ph. D. Thesis, T. M. Bhagalpur University, Bhagalpur, India.

Sumner, F.B. and Lanham, U.N. (1942): Studies on the metabolism of warm and cool spring fishes. Biol. Bull., 82: 313 – 327.

Svehla, G. (1985): *Vogel's Textbook of macro and semi micro qualitative inorganic analysis*. 7th Ed., pp. 48 – 52. Orient Longman Limited, New Delhi.

Svendsen, G.E. (1969): Annotated bibliography on Methylpentynol. Invest. Fish control, No. 32, 1 – 7. U. S. I. B. Sport Fish. Wildl.

Sylvester, J.R. (1975): Factors influencing the efficacy of MS_{222} to striped mullet (*Mugil caphalus*). Aquaculture, 6: 163 – 169.

Sylvester, J.R. and Holland, L.E. (1982): Influence of Temperature, Water Hardness and Stocking Density on MS–222 Response in Three species of Fish. Prog. Fish–Cult., 44(3): 138 – 141.

Szkudlarek, M. and Zakêœ, Z. (1996): Application of Propiscin for total anesthesia of pikeperch (*Stizostedion lucioperca* L.) – Komun. Ryb. 6: 7-8.

Tamura, S.O., Morii, H. and Yuzuriha, M. (1976): Respiration on the amphibious fishes, *Periophthalamus chinensis* in water and on land. J. Exp. Biol., 65: 97 – 107.

Tarazona, J.V., Munoz, M.J., Ortiz, J.A., Nunez, M.O. and Camargo, J.A. (1987): fish mortality due to acute ammonia exposure. Aquacult. and Fish. Manage. 18: 167 – 172.

Thakur, N.K. (1976): On the spawning behaviour of *Clarias batrachus*. Japan J. Ichthyol., 23(3): 178 – 180.

Throat, S.R. and Wagh, S.B. (2001): Oxygen consumption in freshwater Fish, *Channa gachua* exposed to tannery effluent. Indian J. Environ. & Ecoplan. 5(1): 203 – 206.

Thurston, R.V., Phillips, G.R. and Russo, R.C. (1981): Increased toxicity of ammonia to rainbow trout (*Salmo gairdneri*) resulting from reduced concentrations of dissolved oxygen. Can. J. Fish. Aq. Sci. 38: 983–988.

Thurston, R.V., Russo, R.C. and Vinogradov, G.A. (1981): Ammonia toxicity to fishes. Effects of pH on the toxicity of the un-ionized ammonia species. Environ. Sci. Tech., 15: 7–12.

Thurston, R.V., Russo, R.C., Luedtke, R.J, Smith, C.E., Meyn, E.L., Chakoumakos, C., Wang, KC. and Brown, C.J.D. (1984): Chronic toxicity of ammonia to rainbow trout. Trans. Am. Fish. Soc. 113: 56 – 73.

Thurston, R.V., Russo, R.C., Meyn, E.L., Zajdel, R.K. and Smith, C.E. (1986): Chronic toxicity of ammonia to fathead minnows. Trans. Am. Fish. Soc., 115: 196–207.

Thurston, R.V. and Russo, R. C. (1978): Acute toxicity of ammonia and nitrite to cutthroat trout fry. Trans. Am. Fish. Soc. 107: 361–368.

Thurston, R.V. and Russo, R.C. (1983): Acute toxicity of ammonia to rainbow trout (*Salmo gairdneri*). Trans. Am. Fish. Soc. 112: 696–704.

Thurston, R.V., Russo, R.C. and Phillips. G.R. (1983): Acute toxicity of ammonia to fathead minnows (*Pimephales promelas*). Trans. Am. Fish. Soc., 112: 705 – 711.

Tomasso, J. R., Goudie, C. A., Simco, B. A. and Davis, K. B. (1980): Effects of environmental pH and calcium on ammonia toxicity in channel catfish. Trans. Am. Fish. Soc. 109: 229–234.

Tsukuda, N. and Arai, K. (1986): Effect of defatted sardine and Alaska Pollock meat on plasma lipid composition of rat fed with a cholesterol containing diet. Bul. Tokas. Reg. Fish. Res. Lab. 120: 61 – 67.

Uematsu, T. (1971): Social facilitation in feeding behaviour of the guppy. III. Influences of the social facilitation upon the activity, respiration and growth. Japan. J. Ecol., 2: 96 – 103.

Umezawa, S.I. and Kanikawa, M. (1980): Effects of fishes of different species on oxygen consumption of the goby, *Tidentiger oscurus*. Annot. Zool. Japan., 54(2): 93 – 103.

Umezawa, S.I. and Watanabe, H. (1979): Group effect on the respiration of the medaka *Oryzias latipes*. Japan. J. Ichthyol., 26(3): 266 – 272.

Umezawa, S.I. Ueda, H. and Shibayama, H. (1981): Group effect on oxygen consumption and some Observations on the habit of goby, *Tridentiger obscurus*. Annot. Zool. Japan., 54(2): 93 – 103.

Umezawa, S.I., Adachi, S. and Taneda, K. (1983): Group effect on oxygen consumption of the Ayu (*Plecoglossus altivelis*.) in Relation to Growth Stage. Jap. J. Ichthyol., 30(3):261–267.

Vaas, K.F. (1951): Preliminary report on air transport of live fish in scaled cans under oxygen pressure Proc. Indo Pacif. Fish. Concil, India. IPFC/C51/Tech. 46.

Vander, P. and Part, P. (1982): Oxygen and chromium transfer in perfused gills of rainbow trout (*salmo gairdneri*) exposed to hexavalent chromium at two different level. Aquatic Toxicol, 21: 41 – 45.

Varga, I. (1984): Magyar anyahalak egyiptomi halkeltetö telepen (Transportation of brood carp and brood herbivorous fishes from Hungary to Egypt). Halaszat, 30(3):88–89.

Veenstra, R.S., Balon, E.K. and Flegler-Balon, C. (1987): Propanidid, a useful anaesthetic for studying blood circulation in early development of fish – J. Can. Zool. 65: 1286-1289.

Vermeer, G.K. (1988): Evaluation of metomidateasan anesthetic for common snook, *Centropomus undecimalis* – Contributions in Marine Science, Port Aranas TX, 30: 197.

Wajsbrot, N., Gasith, A., Diamant, A. and Popper. D.M. (1993): Chronic toxicity of ammonia to juvenile gilthead seabream *Sparus aurata* and related histopathological effects. J. Fish. Biol. 42: 321–328.

Wang, S.C. and Nagi, S.H. (1964): General organization of central respiratory mechanisms. In *Handbook of Physiology* (Ed. Am. Physiol. Soc., J. Field), I, pp. 487 – 505. Williams & Wilikins, Batimore, Maryland.

Ward, A.J.W. and Webster, M. (2016): *Sociality: the behaviour of group-living animals*. Berlin, Germany: Springer.

Wares II, W.D. and Igram, R. (1979): Oxygen consumption in the fathead minnow (*Pimephales promelas* Rafinesque). I. Effects of weight, Temperature, Group size, Oxygen level and Opercular Movement rate as a function of temperature. Comp. Biochem. Physiol., 62(A): 151-156.

Wassersug, R.J. and Johnson, R.K. (1976): A remarkable pyloric caecum in the ever mannellid genus *Coccorella* with notes on gut structure and function in alepisaurid fishes (Pisces, Mytophiformes) J. Zool. Lond.

Wedemeyer, G. (1970): Stress of anaesthesia with MS_{222} and benzocaine in rainbow trout (*Salmo gairdneri*). J. Fish. Res. Bd. Can., 27: 909 – 914.

Welch, P.S. (1948): *Limnological methods*. McGraw Hill, New York.

Wells, M.A. (1935): Variations in the respiratory metabolism of the Pacific killifish, *Fundulus parvipinnis*, due to size, season and continued constant temperature. Physiol. Zool., 8: 318 – 335.

Wells, M.M. (1913): The resistance of fishes to different concentration and combinations of oxygen and carbon dioxide. Biol. Bul., 5(6): 323 – 361.

Wiebe, A.H., Mc Gavock, A.M., Fuiier, A.M. and Markus, A.H. (1934): The ability of freshwater fish to extract oxygen at different hydrogen ion concentrations. Physiol. Zool., 7: 435 – 448.

Wilson, R.P., Anderson, R.O. and Bloomfield, R.A. (1969): Ammonia toxicity in selected fishes. Comp. Biochem. Physol. 28: 107 – 118.

Winberg, G.G. (1956): Rate of metabolism and food requirements of fishes (Original in Russian). J. Fish. Res. Bd. Can., trans. Ser. no. 194.

Wipfli, M.S., Merritt, R.W. and Taylor, W.W. (1994): Low toxicity of the black fly larvicide *Bacillus thuringiensis* var. *israelensis* to early stages of brook trout (*Salvelinus fontinalis*) brown trout (*Salmo trutta*), and steelhead trout (*Oncorhynchus mykiss*) following direct and indirect exposure. Can. J. Fish. Aq. Sci., 51: 1451–1458.

Woker, H amd Wuhrmann, K. (1950): Die Empfinlichelt Vershiedner Fischarten gengeniiber Ammoniak, Blausaure und Phenest. Revue Suisse de Zoologic 57: 458 – 553.

Wolischlag, D.E. and Juliano, R.O. (1959): Limnol. Oceanogr. (4): 195 – 209.

Wood, E.M. (1956): Urethane as carcinogen. Prog. Fish – Cult., 18: 135 – 136.

World Health Organization (WHO) (1986): Environmental Health Criteria. 54. Ammonia, Geneva.

Woynarowich, E. and Horváth, L. (1980): The artificial propagation of warm-water finfishes - a manual for extension. FAO Fish. Tech. Pap, (201): 138–47.

Writz, P. and Davenport, J. (1976): Increased oxygen consumption in blennies (*Blennius Pholis* L.) exposed to their minor images. J. Fish. Biol., 9: 67–74.

Xiaoqin, Li., Zhenhua Z. and Jing, J.Z. (2017): Research progress in the use and detection techniques of fishery anesthetics. Agricultural Sciences. 45(29):72-73.

Xiaowang, S., Yongping, C. and Tong, L. (2017): and so on. Study on the Anesthesia Effect and Elimination Law of MS-222 on *Cynoglossus semilaevis*. Aquatic Science and Technology Information. 44(2):87-90.

Yamamoto, K.I. (1991a): Relationship of respiration to body weight in *Cyprinus Carpio* under resting and normoxic corditions. Comp. Biochem. Physiol., 100A: 113 – 116.

Yamamoto, K.I. (1991b): Relationship of respiration to body weight in the tilapia *Oreochromis nitolicus* under resting and normoxic conditions. Comp. Biochem. Physiol., 103A(1): 81 – 83.

Yatao, D., Zhihe, W. and Linlin, W. (2019): Study on anesthesia, preservation and transportation effects of MS-222 on bream. Aquatic Science, 38(3):296-304.

Yinlong, Y., Yonghai, S. and Haiming, Z. (2016): Anesthetic effect of MS-222 and eugenol on juvenile *coilianasus.* Journal of Shanghai Ocean University, 25(2):177-182.

Zeuthen, E. (1947): Body size and metabolic rate in the animal kingdom: Cr. Trav. Lab. Carlberg (Serchim). 26: 17 – 61.

Zeuthen, E. (1953): Oxygen uptake as related to body size in organisms. Q Rev Biol 28: 1 – 12.

Zhang Y, Matthew J.H. Gilbert and Anthony P.F. (2019): Finding the peak of dynamic oxygen uptake during fatiguing exercise in fish. Journal of Experimental Biology. 222, jeb196568. doi:10.1242/jeb.196568.

Author Index

D

E

F

G

H

I

J

N

O

P

R

S

T

T

V

W

Subject Index

E

F

G

H

I

J

L

M

U

V

W